The Engineer and Society

Other Macmillan titles of related interest

The Management of Manufacturing Systems
J.D. Radford and D.B. Richardson

Production Engineering Technology, second edition
J.D. Radford and D.B. Richardson

The Engineer and Society
An Introduction to
Engineering Management

J.D. Radford
B.Sc. (Eng.), F.I.Prod.E., M.I. Mech.E.

MACMILLAN

© J.D. Radford 1984

First published by
Higher and Further Education Division
MACMILLAN PUBLISHERS LTD
London and Basingstoke
Companies and representatives
throughout the world

ISBN 0 333 35815 5
ISBN 0 333 35816 3 (pbk)

Printed in Kong Kong

British Library Cataloguing in Publication Data

Radford, Dennis
    The engineer and society.
    1. Engineering——Great Britain——Management
    I. Title
    620'.0068   TA190

    ISBN 0-333-35815-5 ✓
    ISBN 0-333-35816-3 Pbk

Contents

Preface

This book has been written because there is no other single text that attempts to help students who are preparing for examinations in 'The Engineer in Society'. Also it is hoped that the book will be of use to students following a variety of engineering courses in which there is some management content.

The text is based on lecture notes used by the author in teaching 'The Engineer in Society' to students reading for degrees in mechanical engineering.

Many may feel that this book is too brief, but its brevity is intentional and designed to keep the price within the book-buying budgets of most students.

Engineers have been referred to in the masculine gender. This was done to avoid the form him/her or he/she; it in no way implies a refusal to recognise the growing importance of women in the profession.

J.D. Radford

Acknowledgements

I should like to express my appreciation to all those who have assisted in the preparation of this book; in particular my colleagues Mr Don Richardson for his helpful criticism and Mrs Brenda Foster and Miss Karen Harman for their help in typing the manuscript.

1 Development of Engineering Technology

1.1 PRE-EMINENCE AND DECLINE

The eighteenth and early nineteenth centuries were times of great achievement for British engineering. The important contribution made by Britain during this period is shown in figure 1.1. This good start enabled us to export the ideas and products of the Industrial Revolution to the rest of the world. During the latter part of the Victorian period we became complacent and our position was being challenged by the United States and Germany. The twentieth century has seen a decline in our share of world trade in manufactured goods and the productivity of our manufacturing industries has fallen below that of our major competitors. British inventive genius still thrives but sections of our industry are badly managed and suffer from outdated trade union attitudes.

In a single chapter it is possible only to provide a brief and incomplete view of the development of engineering since the beginning of the Industrial Revolution. A similarly brief treatment has been given to the development of the engineering institutions and engineering education. The first part of this chapter is concerned with the development of engineering technology and is grouped under the headings of engineering materials, steam-power, machine tools, transport, and electricity.

1.2 ENGINEERING MATERIALS

1.2.1 Iron and Steel

Cast iron and steel continue to be the most commonly used engineering materials; they are relatively cheap and their versatility makes them suitable for a wide range of applications.

Iron-making was brought to Britain by Celtic immigrants in about 450 BC. Primitive blast furnaces were in operation by the end of the fifteenth century. These were charged with iron ore, charcoal, and limestone; it required six tonnes of charcoal for each tonne of iron produced. The industry was centred in the forests of the Sussex Weald and the Forest of Dean. The steadily increasing demand for iron in the sixteenth and seventeenth centuries denuded the forests and created a shortage of charcoal.

Iron-making with Coke

Pig-iron was first smelted with coal instead of charcoal early in the seventeenth century. However the major breakthrough came in 1709 when Abraham Darby I (1677-1717), a Quaker from Coalbrookdale in Shropshire, used coke in the manufacture of pig-iron. Darby's son Abraham Darby II (1711-63) fully established his father's process in 1750; he used haematite ores to produce iron of improved quality. The Darbys' ideas were gradually adopted by other iron masters and

British		Non-British
Savery's atmospheric engine	1700	
Darby's iron-making with coke Newcomen's atmospheric engine		
	1750	
Watt's pumping engine		Cugnot's steam road engine
Cort's wrought iron Watt's rotative engine		Montgolfier's hot air balloon Charles's hydrogen balloon
Bramah's hydraulic press Maudslay's screw-cutting lathe	1800	Volta's battery
Trevithick's steam locomotive		Fulton's steamship
Faraday's electric motor Stephenson's first public railway		Whitney's milling machine
Cooke and Wheatstone's telegraph		Morse's telegraph
	1850	Giffard's airship
Bessemer's converter Siemens's open hearth furnace Parkes's Parkesine plastic		Lenoir's gas engine
		Beau de Rochas's four-stroke cycle Bell's telephone Edison's electric light bulb
Parsons's steam turbine Ackroyd Stuart's oil engine Dunlop's pneumatic tyre		Daimler's petrol engine Benz's motor car
		Diesel's engine
	1900	Marconi's radio transmission Zeppelin's airship
Fleming's diode		Wright brothers' first flight De Forest's triode Baekeland's Bakelite plastic
		Krupp's tungsten carbide
Whittle's turbo-jet patent		Zworykin's television camera
		Carother's nylon Nuclear reactor Transistors
	1950	Electronic computer

Figure 1.1   Some major engineering developments since 1698

in consequence the cost of iron was reduced and its output increased.
The use of coke in iron-smelting concentrated the production of iron
in coal-rich areas such as South Wales, South Yorkshire and the Black
Country.

Wrought Iron
The work of Henry Cort (1740-1800) was of comparable importance to
that of the Darbys. When working as a contractor to the Navy he
found that iron bars were being imported from Russia and Sweden as
the British product was inferior. He decided to try to capture this
trade and set up a foundry near Fareham in Hampshire. He remelted
pig-iron in a reverberatory furnace using coal as the fuel. It will
be seen from figure 1.2 that the coal and the iron were separated so
that contamination of the iron by sulphur from the coal was avoided.
The molten iron was puddled, a process in which the iron was stirred
with long rods to oxidise the carbon and impurities. After removal
from the furnace, the iron was hammered to expel slag then passed
through a grooved rolling mill to produce bar. Cort's discoveries of
1783-84 were patented and taken up by iron-masters throughout the
country thus increasing the production of wrought iron and improving
its quality.

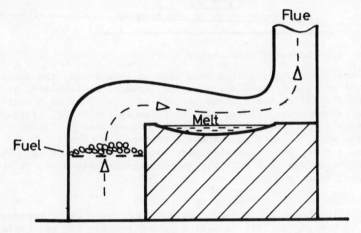

Figure 1.2 Cort's puddling furnace

Steel-making with the Bessemer Converter
Sir Henry Bessemer (1813-98) was an inventor of great versatility
whose major discovery was to produce steel cheaply. Mild steel became an alternative to wrought iron and eventually replaced it as an
engineering material. In Britain by 1850, most of the annual output
of two-and-a-half million tons of iron was converted into wrought
iron by the costly process of puddling. James Nasmyth in 1854 had
patented the idea of removing unwanted carbon from iron by blowing
steam through it. The process proved unreliable and was not developed.
   Bessemer had noticed that iron bars became decarburised when left
in a draught of air at the edge of a furnace. This observation led in
1856 to Bessemer's patent for the manufacture of steel by blowing air
through molten pig-iron in a Bessemer converter as shown in figure
1.3. Two early difficulties were experienced. One was hot shortness

(cracking of the metal when worked hot); this was overcome by using non-phosphoric iron ores. The other, the occlusion of excess oxygen, was overcome by Robert Forester Mushet (1811-91), the owner of a small ironworks in Gloucestershire, who added manganese to the molten iron.

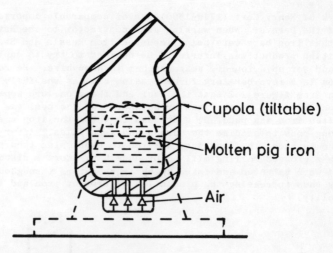

Figure 1.3 Bessemer's converter

As most British iron ores were high in phosphorus, the Bessemer process was mainly used abroad under licence, particularly in the United States. However, from 1871 to 1875, a clerk in a metropolitan magistrates' court, Sidney Gilchrist-Thomas (1850-85), assisted by his cousin, a chemist in an ironworks, found that by lining a blast furnace with dolomite (a carbonate of lime and magnesium) and by adding a flux, the phosphates were absorbed into the slag. The resulting iron was then suitable for conversion into steel by this Bessemer process.

The Siemens-Martin Method of Steel-making
Sir William Siemens (1823-83) trained in Germany as an engineer. He came to England in 1843 to sell a patent process for metal-plating invented by his brother Werner. William settled, married, and became a British subject.

His most important invention was patented in 1856; it was the open-hearth furnace, which he developed with another brother, Frederick. The lower part of this furnace was a honeycomb of bricks divided into two halves, one for incoming, the other for outgoing gases as shown in figure 1.4. When the outgoing gases had heated the bricks, the gas flow was reversed to pre-heat the incoming air. These alternations of flow were maintained during the operation of the furnace. Siemens's patent of 1861 was for a gas producer supplying an open-hearth furnace; the gas and the air were separately pre-heated then brought together at the edge of the furnace and ignited. The gas producer used cheap coal and the heat interchanger in the furnace gave good fuel economy and a high furnace temperature.

The original use of the open-hearth furnace was in the manufacture of glass at Chance Brothers in Smethwick. Its first use in the steel industry was to melt steel for castings. The two French brothers

Émile and Pierre Martin patented in 1865 the most important use of
Siemens's furnace. This was to make steel by melting a mixture of
pig-iron, scrap steel, and limestone. By 1900 the Siemens-Martin
process was the most important method of steel-making in Britain.

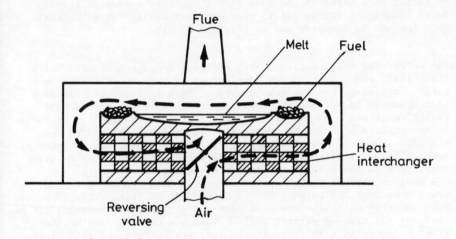

Figure 1.4 Siemen's original reverberatory furnace

Recent Developments in Steel-making
In bulk steel-making the open-hearth process has been almost completely replaced by the basic oxygen furnace process in which oxygen
is blown on molten iron to remove impurities. This process has been
made viable by low-cost supplies of liquid oxygen that have become
available since the Second World War.

  Continuous casting, which was introduced in the 1950s, is being
increasingly used in steel-making. Molten steel from the furnace is
poured into an open-ended water-cooled mould from which white-hot bar
emerges; the bar is then rolled. This process saves power, material,
and time as compared with discontinuous large-ingot production, where
the ingot is first cast then removed from the mould and subsequently
rolled.

1.2.2 Alloy Steels
Manganese Steel
This alloy was discovered in 1882 by Sir Robert Hadfield (1858-1940);
it consisted of steel containing 1 per cent carbon to which 13 per
cent manganese had been added. If quenched, after heating to 1000°C,
manganese steel is exceptionally resistant to abrasion. It was used
for railway points and for the jaws of rock-crushing machinery.

Nickel Steel
Another early alloy was nickel steel, the nickel being added to a
carbon steel to increase its strength. This group of alloys was
later improved by adding chromium and a small quantity of molybdenum.

Stainless Steel
Stainless or 'rustless' steel was discovered in Sheffield in 1913 by

Harry Brearley who beat the German and American inventors by a short head. Brearley had been experimenting with different types of steel for gun barrels; some months afterwards he noticed that one of his discarded test-pieces, containing 14 per cent chromium, had not rusted. This observation led to the introduction of stainless steel. By adding more chromium and some nickel, greater corrosion resistance, toughness, and ductility were obtained making the alloy suitable for use in furnaces and in chemical plants.

Tool Materials

The celebrated American industrial engineer Frederick Winslow Taylor (1856-1915) was active in the development of high-speed steel, an alloy containing chromium and tungsten. This material had already been produced in England by Mushet in 1872. However, around 1895 Taylor developed a method of heat treatment that enabled its hardness to be retained at high temperatures. High-speed steel caused considerable interest when exhibited at the Paris Exhibition of 1900; it revolutionised metal-cutting, enabling cutting speeds to be trebled compared with those available with plain carbon tool steels.

A further major advance in tool materials occurred in 1928 when the Krupp's company exhibited tungsten carbide at Leipzig. This is a very hard material which, in powdered form and sintered with cobalt, can be used for the tips of cutting tools. Tungsten carbide retains its great hardness at high temperatures and this enabled a further trebling of cutting speeds to be achieved. In 1969 coated tungsten carbide became available and further substantially improved cutting performance. The coating is a layer of ultra-hard materials that are vapour deposited; one type of tool is coated first with titanium carbide, then aluminium oxide, and finally titanium nitride.

1.2.3 Aluminium and its Alloys

Although of relatively late introduction, aluminium and its numerous alloys are of great importance as engineering materials. The density of aluminium is one-third that of steel, it has good electrical and thermal conductivity, and does not corrode easily. Pure aluminium is weak but its alloys can be as strong as medium carbon steel. It is readily shaped by rolling, extrusion, and casting and can be machined at high cutting speeds.

Production of Aluminium

Owing to production difficulties, aluminium was a precious metal until about 1890. The breakthrough came in 1886 when the American Charles Hall and the Frenchman Paul Heroult independently discovered how aluminium could be economically produced. It is a coincidence that both men were born in 1863 and died in 1914. They found that when aluminium oxide was mixed with the mineral cryolite (sodium aluminium fluoride), the mixture would melt at 1000°C, thereby allowing the aluminium oxide to be split by electrolysis into aluminium and oxygen.

Aluminium Alloys

Duralumin was one of the important early alloys of aluminium. It was discovered, more or less by accident, in 1906 by the German metallurgist Alfred Wilm. He had been experimenting in the Durener metal works to find the effect of heat treatment on aluminium alloys that contained copper. Some days later, on checking the hardness of a specimen containing 3.5 per cent copper and 0.5 per cent magnesium, he found to his surprise that its hardness had increased considerably. This led to further tests and to the discovery of a precipitation-hardening alloy of strength similar to that of mild steel. Duralumin was used extensively in the construction of Zeppelins and later in aeroplanes.

Many casting alloys of aluminium have been developed. An early alloy was one that contained about 10 per cent copper. Another of more recent introduction contained about 12 per cent silicon; this had a higher strength and better corrosion resistance than the earlier casting alloys.

1.2.4 Copper and its Alloys

In Britain, copper and tin occur together in ores that, when smelted, produce bronze - a hard alloy. Bronze is an ancient metal and articles dating back to 1800 BC have been found in Britain. In the first part of the nineteenth century, most of the world's copper requirement was smelted in South Wales, the ores coming from Cornwall and Spain.

About half of the copper now produced is used to conduct electricity. The remainder is used in alloys with wide engineering applications. The best known of these is brass, the properties of which vary considerably with the amount of zinc present. There are records of brass being cast in London in 1693.

Phosphor bronze is a well-known copper alloy that contains tin and phosphorus. Other bronzes, such as silicon and aluminium bronze, contain no tin, the term 'bronze' being used to indicate a superior material when compared with brass. Leaded bronze, which contains up to 30 per cent lead and small quantities of strengthening elements, is an important high-speed bearing alloy.

1.2.5 Titanium and its Alloys

The production of titanium is the most recent major advance in non-ferrous technology. Titanium is about half-way between steel and aluminium in density but nearer to steel in strength.

A Cornish clergyman, the Rev. William Gregor (1762-1817), recognised titanium oxide as the oxide of an unknown metal as early as 1789. It was not until 1905 that the pure metal was produced and then only on a laboratory scale. In 1937, Wilhelm Kroll from Luxemburg developed a suitable method of extraction in which he produced titanium sponge by magnesium reduction of titanium tetrachloride. In the following year he successfully melted the metal in a water-cooled arc furnace; a tungsten electrode was used and melting was in an argon atmosphere. Apart from detail changes, Kroll's methods of extraction and melting are still in use.

Titanium and its alloys were not produced in significant quantities until about 1950. Since then there has been a rapidly growing market despite the high cost of the product. Owing to its lightness and strength, titanium has found a ready market in the aircraft industry where it is used for both airframe construction and in engines. The chemical industry is another major market where titanium is valuable as it is able to withstand the highly aggressive conditions found in chemical plant.

1.2.6 Nickel and its Alloys

Nickel was first isolated in 1751 by the Swede Axel F. Cronstedt

(1722-65). Its use as an alloying element in steel has been mentioned in section 1.2.2.

Although the major use of nickel is in alloys, pure nickel is
extensively used in plating. An early alloy of nickel was Monel
metal. This was introduced by Ambrose Monell in 1905; it was smelted
directly from ores containing two parts of nickel and one part of
copper and mined in Canada by the International Nickel Company.
Monel metal has good mechanical properties and corrosion resistance;
its applications include turbine-blading and pump parts.

Another group of alloys are those suitable for high-temperature
applications. The original alloy in this group was patented by
A.L. Marsh in a British patent of 1906. It consisted of 80 per cent
nickel and 20 per cent chromium and was used for the manufacture of
electric heating elements. In 1939 Whittle required a creep-resist-
ant alloy for his gas turbine engines. This led to the development
of a large number of wrought and casting alloys marketed by the
Henry Wiggin Company under the trademark of Nimonic. These and other
similar alloys are suitable for applications in gas turbines where
prolonged resistance to temperatures of up to 1000°C is required.

1.2.7 Plastics
A plastic can be defined as an organic material that may be shaped
by the application of heat and pressure and will retain its new form
after shaping. The use of plastics has taken off since the 1940s.
They are generally lighter and more easily shaped than metals although they are weaker and less able to operate at elevated temperatures.

The history of plastics is relatively recent. The first plastic
was invented in 1862 by the Englishman Alexander Parkes (1813-90) and
was known as Parkesine. The first commercially successful plastic,
celluloid, was introduced in 1868.

The earliest plastic of major industrial importance was Bakelite.
This was introduced in 1910 by the Belgian-born American Leo Hendrik
Baekeland (1863-1944) who was seeking an electrical insulating
material to replace shellac and ebonite. Bakelite was based on
phenolic resin which could be moulded to shape from a powder, under
the effect of heat and pressure. In the 1930s urea formaldehyde
resin was introduced; this enabled the dark colours of phenolic resin
to be replaced by a material that was translucent and could be
attractively coloured.

Although much of the basic work on polymerisation had been done by
German chemists, it was in the United States in 1928 that a satisfactory production method for polyvinyl choride was developed. In
the same year Wallace Carothers (1896-1937) started his work on
polymerisation at the Dupont Company which culminated in the discovery of nylon in 1935.

Low-density polyethylene was introduced in 1933 with the high-
density version following in 1955. In 1936, polystyrene first
appeared followed in 1940 by melamine plastics; of the newer materials, polypropylene was introduced in 1957 and polycarbonate in 1958.

1.2.8 Composite Materials
Plastics may be strengthened by the introduction of a reinforcing
material. In the 1950s interest in stronger plastic materials came
from boat-builders as a replacement for wood, from aircraft manufacturers who were seeking to save weight, and from small-quantity car

manufacturers who were looking for an economical method of manufacturing bodies.

Glass Fibre-reinforced Plastic

This widely used composite normally has randomly orientated glass
fibres in a matrix of epoxy resin. The presence of glass fibres can
increase the stiffness of the resin by a factor of fifteen.

In the 1950s when the Rolls-Royce Company first considered ways of
reducing the weight of gas turbine engines, they investigated factors
affecting the mechanical properties of glass-reinforced plastics.
Strength could be increased by attention to fibre direction and by
minimising the number of gas bubbles in the resin. Also it was possible to increase the maximum working temperature from 200°C to 240°C
by replacing epoxy resin by polyamide resin.

Carbon Fibre Composites

In the early 1960s two new reinforcing materials, boron fibre and
carbon fibre, became available; both these had much higher stiffness
than glass fibre. By 1968, carbon fibre composites were being used
for the compressor blades in the engines of some types of passenger
aircraft. Since then there has been a gradual improvement in carbon
fibre composite production and an increase in their usage, both in
engines and elsewhere in aircraft. In performance, carbon fibre
composites are generally comparable with their metal alternatives
although they exhibit considerably more anisotropy. The weight-
saving advantage of composites must, however, be set against their
higher production costs.

A second generation of composites is now being developed, for
example, the use of metal or ceramic matrices reinforced by fibres.

1.3 STEAM POWER

Before man harnessed steam-power he depended on animals, water, and
the wind to do work for him. Animals were limited in their working
capacity, water-power was limited in its availability, both geograph-
ically and seasonally, and wind-power required suitable locations
and, of course, the right weather conditions.

Newcomen's atmospheric engine provided mine-owners with an opportunity of draining water from their mines. Watt's rotative steam-
engine provided a greatly improved method of driving machines in
factories and mills. Trevithick's high-pressure steam-engine enabled
him to develop the steam locomotive.

1.3.1 The Atmospheric Engine

In 1698 Captain Thomas Savery (1650-1715), a military engineer, invented and patented the first atmospheric engine. This 'fire
engine', as it was called, operated by filling a vessel with steam,
then pouring water over the surface of the vessel thereby causing the
steam to condense and suck water up a pipe into the vessel. See
figure 1.5. When the vessel had filled with water, the valve to the
suction pipe was closed and steam was admitted to the vessel to blow
the water up a pipe and into the air. As the total lift was only
about 20 metres, Savery's engine had limited use for mine drainage.

Thomas Newcomen (1663-1729), a Dartmouth blacksmith, designed a
more successful pumping engine. He condensed steam in a large brass
cylinder, the action of this condensing steam pulling down a

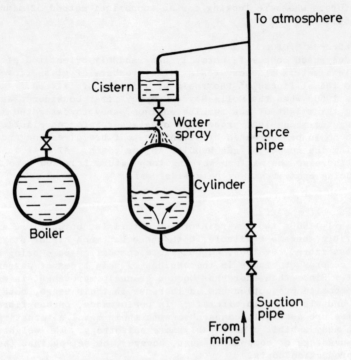

Figure 1.5 Savery's atmospheric engine

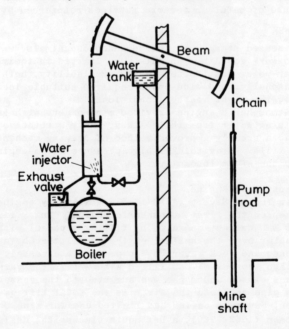

Figure 1.6 Newcomen's atmospheric engine

piston in the cylinder. The piston rod was attached to one end of a
rocking beam; the other end of this beam was attached to a pump as
shown in figure 1.6. As Savery's patent covered all 'fire engines',
Newcomen had to take Savery into partnership. Newcomen's first
engine was erected in 1712 near Dudley Castle. These engines had a
hearty appetite for fuel, their overall thermal efficiency being about
0.5 per cent. However, when used for pumping in coal-mines they
could be fired with unsaleable coal. Despite the high cost of operating atmospheric engines in areas remote from coal-mines, there were
75 working in the Cornish tin-mines by 1777.

1.3.2. The Steam-engine

James Watt (1736-1819) was working in Glasgow as 'Mathematical
Instrument Maker to the University' when he was surprised by the very
large amount of steam needed to operate a model of a Newcomen engine
that he was repairing. He redesigned Newcomen's engine to produce
the first true steam-engine and in 1769 was granted a patent for 'A
new Method of Lessening the Consumption of Steam and Fuel in Fire
Engines'. He made these savings by designing a condenser separate
from the cylinder, keeping the cylinder hot by means of a steam-
jacket, and preventing steam escaping past the piston rod by invent-
ing a stuffing box as shown in figure 1.7. These improvements increased the overall thermal efficiency of the steam-engine to about
2.5 per cent.

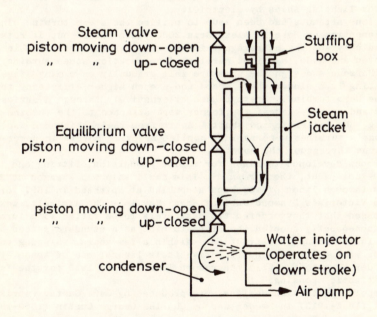

Figure 1.7 Principle of operation of Watt's steam-engine

In 1775 Watt went into partnership with Matthew Boulton (1728-1809)
at the Soho factory near Birmingham. One year later the first of
Watt's steam-engines had been manufactured and installed. By 1780,

40 pumping engines had been sold, many of these replacing less efficient Newcomen engines and enabling deeper mines to be sunk.

Boulton and Watt next turned their attention to the manufacture of
a steam-engine that could produce rotary motion and hence power mills
and factories. In 1782 Watt patented a double-acting beam engine in
which the piston was driven on both the up and down strokes. To enable the piston to both push and pull the beam, Watt designed a
parallel-motion linkage to connect piston rod and beam. At the other
end of the beam there was a connecting rod which drove the fly-wheel
shaft by means of a sun and planet gear. A centrifugal governor was
also driven from the fly-wheel shaft which enabled the engine to maintain a constant speed by controlling the steam flow to the cylinder.
The first rotative engine was supplied in 1783; by 1800, 321 engines
had been installed, of which 84 had been supplied to cotton-mills.
Watt's rotative engine revolutionised methods of manufacture by taking production out of homes and into factories and mills where machines could be driven by belts and shafts from a steam-engine.

1.3.3 The Steam Turbine

Sir Charles Parsons (1854-1931) was the sixth son of the third Earl
of Rosse. He was educated at Trinity College, Dublin and St. John's
College, Cambridge. After university he served a four-year premium
apprenticeship at the Elswick works of the W.G. Armstrong Company.
Subsequently, while working as a junior partner in the Clarke Chapman
Company, he became interested in using steam turbines to drive dyna-
mos for lighting ships by electricity.

Various attempts had been made to utilise the steam turbine since
its invention by Hero of Alexandria 2000 years before but it rotated
too quickly and was extravagant in the use of steam. Parsons in 1884
patented and made the first parallel-flow reaction steam turbine.

In Parsons's design the pressure fell gradually over many stages,
resulting in a slower shaft speed and a much higher efficiency than
in the Hero turbine. Steam passed alternatively through a series of
fixed and moving blades; the former were attached to the turbine
casing, the latter to the shaft. Apart from their shipboard use,
Parsons's turbines were rapidly applied as prime movers in power-
stations throughout the country.

Parsons developed turbines for marine propulsion, fitting one to
a 100-foot yacht, the *Turbinia*. This small ship was steamed at 34½
knots through lines of warships assembled at Spithead in 1897 for
Queen Victoria's Diamond Jubilee Naval Review. The Admiralty was so
impressed that they ordered two destroyers to be fitted with turbines
and subsequently adopted the steam turbine as a standard method of
propulsion for new construction. Within a few years, shipping companies were using turbines to drive their liners; one of these, the
Cunarder *Mauretania*, held the record from 1907 to 1929 for the fast-
est Atlantic crossing.

Competitive turbine designs were produced by Carl Gustav Patrik de
Laval (1845-1913) in Sweden and by Charles George Curtis (1860-1953)
in the United States. Both of these were impulse turbines where the
expansion of the steam occurred in the nozzles of stationary blades.
Parsons's turbine was of the reaction type in which most of the steam
expansion took place within the moving blades and motion resulted
from reaction caused by the steam as it left the blades.

De Laval's turbine was a single-stage machine which operated at
very high rotational speeds. His first steam turbine was little more

than a cased Hero turbine designed to operate a cream-separator he had designed. A more efficient turbine was designed by de Laval in 1889 in which the steam struck the blades at a high velocity and produced shaft speeds of up to 30,000 revolutions per minute. There were difficulties in gearing down the turbine speeds for electrical generators but a number of de Laval turbines of limited capacity went into commercial use.

Curtis patented his multi-stage impulse turbine in 1895; it was successful and used for electrical power generation. The standard power-station turbine is now the impulse-reaction type which incorporates the advantages of both Parsons's and Curtis's designs.

Although steam-power has largely disappeared in other applications, it is still supreme in the power-station. Boiler, turbine, and generator efficiencies have increased considerably during the twentieth century; it now requires less than half a kilogram of coal to generate one kilowatt-hour of electricity, compared with about 3 kilograms in 1900.

1.4 MACHINE TOOLS

Without machine tools it would have been impossible to have achieved nineteenth century industrialisation or to have satisfied the mass markets of the twentieth century. From the crude and limited types of machine tool available at the start of the Industrial Revolution, a comprehensive range of accurate general and special-purpose machines has been developed. Rates of production and accuracy have increased, skill levels reduced, and some types of machine have been made automatic in their operation.

At the beginning of the Industrial Revolution the principal machine tools were boring machines, lathes, and mechanical hammers, with a number of small machines developed for makers of instruments and clocks.

1.4.1 Boring Machines
A greatly improved boring machine was invented in 1775 by the ironmaster and industrialist John Wilkinson (1728-1808). This machine was fitted with a hollow boring bar which was rigidly supported at each end; the disc that carried the cutters was rotated by the boring bar and was moved along it by a separate mechanism.

Although Wilkinson designed his machine to bore cannon cast in his works, it was available just in time to bore the second cylinder for Watt's experimental steam-engine after the first cylinder had failed.

1.4.2 The Lathe
Primitive lathes were originally used by the Greeks about 1000 BC. An early development was the pole lathe; here rotation was obtained by a string wrapped round the workpiece, the lower end of the string was attached to a treadle, while the upper end was tied to a flexible pole. This arrangement rotated the work backwards and forwards, producing intermittent cutting. Later developments stemmed from Leonardo da Vinci's (1452-1519) design for a lathe with fixed and running centres, which was rotated by a treadle and crank. Clockmakers' lathes of all-metal construction were introduced in France around 1740, while the first practical screw-cutting machine was produced in the 1770s by the British instrument maker Jesse Ramsden (1735-1800).

The Industrial Screw-cutting Lathe

Henry Maudslay (1771-1831) was both a highly skilled craftsman and a first-class creative engineer. He had worked for the prolific inventor Joseph Bramah (1749-1814) who, among other things, invented an 'unpickable' lock, an improved water-closet, a beer engine, and the hydraulic press. When Maudslay left Bramah in 1797 he opened his own business in London, off Oxford Street, firstly in Wells Street and then in Margaret Street where he remained until 1810.

Maudslay's first screw-cutting lathe was made around 1797. His achievement was that he synthesised the important ideas of other eighteenth century lathe designers into a substantial and accurate lathe, which became the prototype of the modern industrial lathe. Between 1810 and 1825 Maudslay's company sold a large number of lathes to British manufacturers, as well as using them in his own engineering business, Maudslay Sons and Field of Lambeth.

Richard Roberts (1789-1864), one of Maudslay's pupils, introduced a number of new features on his lathe of 1817 including the back gear and a four-step pulley, to give eight spindle speeds. Another pupil of Maudslay was the famous British engineer and industrialist Sir Joseph Whitworth (1803-87). He started his own business in Manchester in 1833. By 1850 he was the world's leading manufacturer of machine tools. Whitworth introduced the box design for the basic structure of the lathe, thereby increasing its rigidity. In 1835 he patented the use of a half-nut to engage the lead screw to the carriage and an automatic feed for the cross slide. Whitworth's major contribution to the development of the lathe was as the manufacturer of accurate, well-designed, and reasonably priced lathes.

After 1850 the lead that Britain had taken in lathe development was lost to manufacturers in the United States and later to those in Germany. The centre lathe is employed today for one-off production work and for tool-making. In the form of a computer-numerically controlled lathe it is used for the production of small batches of repetition work.

Turret and Automatic Lathes

The Americans concentrated on making the lathe more productive; Stephen Fitch introduced the turret lathe in 1845. This machine, which used pre-set tools and stops, enabled semi-skilled operators to produce batches of turned parts rapidly.

Automatic lathes came next. Once set, these machines operated automatically until the material ran out. Material was usually supplied in bar form, although parts could be fed from a magazine. Automatic lathes were used to manufacture parts where the batch sizes were large. Single-spindle automatic lathes were introduced in 1872, followed in 1893 by multi-spindle automatics.

1.4.3 Grinding

Grinding fills an important gap in the range of machining processes. It can produce parts to closer tolerances and with smoother surfaces than is possible with milling or turning; it can also cut materials that are too hard to mill or turn.

Grinding Machines

Early machines consisted of a large sandstone wheel which was turned by a handcrank. The earliest true grinding machine was designed by Leonardo da Vinci around 1500. This machine was the first to hold

and to guide the work against the wheel. The surface grinding machine was invented in 1831 by the American, William Stone; three years
later his compatriot James Wheaton developed the first universal
grinding machine. Although grinding-machine development continued in
the United States and Europe, progress was severely restricted because of inadequate grinding wheels.

Grinding Wheels

Apart from natural sandstone wheels, nineteenth-century grinding
wheels were of two types. The first was a cast iron disc with a
layer of abrasive grits stuck by glue to its periphery. The other
was a solid wheel made from abrasive grits bonded together by a
second material. A number of bonds were developed; two of the more
successful were vulcanised rubber and silicate. These were awarded
British patents in 1857 and 1859 respectively.

Engineers had to wait until the end of the nineteenth century before really satisfactory grinding wheels were available. The first
major step forward was the introduction of the vitrified grinding
wheel in 1877. This wheel was made by firing at high temperature a
mixture of clay, feldspar and emery. The second and final advance
also came from the United States with the production of synthetic
abrasives; silicon carbide was available in 1896 and aluminium oxide
a few years later. These new materials were more consistent and
cheaper than the natural abrasives emery (an impure form of aluminium
oxide containing large amounts of iron oxide) and corundum (an almost
pure form of aluminium oxide).

Since 1900, two relatively expensive abrasive materials have been
introduced specially to cut hard materials. The first of these,
diamond, was originally available in a resinoid-bonded wheel made by
the Norton Company in 1934. Very hard materials such as tungsten
carbide are machined with diamond wheels. The discovery of cubic
boron nitride was announced in 1957 by the General Electric Company
of America. This synthetic compound, which was previously unknown
in nature, is almost as hard as diamond but is much more chemically
inert; its main use is in grinding tool steels.

Grinding Developments

Grinding machines have been made more accurate and productive.
Specialised machines such as centreless grinders, thread grinders,
and gear grinders have been introduced.

High-speed grinding in which cutting speeds have been trebled from
the conventional 30 metres per second is being examined. Although
this process will significantly increase the rate of metal-removal,
there are safety problems to be resolved.

1.4.4 Milling

The first milling machine, as distinct from the gear-cutting machines
used by clock-makers, was one made by Eli Whitney (1765-1825) in about
1820 for use in his Connecticut gun factory. In 1848 the American
Frederick Howe (1822-91) designed the first commercially available
plain horizontal milling machine. A more rugged and compact version
of this machine was made by the Lincoln Company in 1855; large quantities of these machines were sold in the United States and Europe.

The first universal milling machine was invented in 1861 by Joseph
Rogers Brown (1800-76), of the Brown and Sharpe Company, in order to
machine the spiral flutes on the newly invented twist drills. This

machine was of the knee and column type and incorporated the essential features of the modern horizontal milling machine. An improvement in milling-cutter design was made in 1864 when Brown patented
his formed milling cutter; this could be rapidly sharpened without
changing the tooth form by grinding the face of each tooth.

The vertical-spindle milling machine first appeared in an American
patent of 1862. British and continental engineers made considerable
contributions to the development of this important type of machine
tool.

During the twentieth century the milling machine has been given
greater rigidity and power. Backlash eliminators have enabled parts
to be milled at each end of the table, thereby increasing production.
Specialised milling machines such as plano and profile milling machines have been developed. Cutter design has been improved and their
life has been increased by the use of tungsten carbide tooth inserts.

In 1953 the Massachusetts Institute of Technology developed a tape-
controlled milling machine which was the precursor of the numerically
controlled and eventually the computer-numerically controlled milling
machine. From the numerically controlled milling machine has been
developed the machining centre with its automatic tool-changing
facility.

1.4.5 Metal-forming

Space does not permit anything but a brief reference to metal-forming.
The invention of the hydraulic press by Joseph Bramah in 1795 and the
steam hammer by James Nasmyth (1808-90) in 1842 provided the means of
achieving the heavy primary deformation of hot metal. Although the
steam hammer is all but extinct, the hydraulic press has been developed for a wide range of deformation operations, where it complements
the mechanical press.

Presswork

The nineteenth century saw a vast growth in the sale of products containing parts pressed from sheet metal. These were produced in
manually or power-operated presses. Pressed parts are often made by
a series of separate operations and successful efforts have been made
to complete the whole of the press operations in a single machine.
Small parts can be made on progressive tools; in these tools the
partly finished component is moved forward stage-by-stage still
attached to the strip until cropped off as a completed part. With
larger parts, a transfer press can be used; here the part is moved
progressively forward from tool to tool in a single long-bedded press.

Cold Extrusion

The manufacture of parts from solid steel billets without pre-heat is
now an important metal-forming process. Not only is the material
utilisation high but the mechanical properties of the part are improved due to work-hardening. Although the process was envisaged in
the nineteenth century, it was not until 1934 in Germany that the
cold extrusion of steel was achieved. The Germans kept the process
a military secret because of its value in munitions production. However, in 1945 details of the method of billet-lubrication, which was
the key to successful cold extrusion, were generally known.

1.5 TRANSPORT

In 1830, Britain had a network of canals and metalled roads designed

around the motive power of the horse. Overseas trade was being
carried by sailing-ships that had changed little in their design for
centuries. Fifty years later almost every town and many villages had
their own railway station and steamships travelled safely and regularly on the sea routes of the world. At the end of the nineteenth
century self-propelled passenger and goods vehicles started to appear
on our roads and in 1903 man first flew in an aeroplane. The twentieth century has seen the birth and growth of air transport and a
vast expansion in road transport. The first casualty was horse
transport followed by the railways and ocean liners.

1.5.1 The Railways

In 1800 the brilliant Cornish engineer Richard Trevithick (1771-1833)
built a compact stationary engine using a steam pressure of three
atmospheres, compared with about half an atmosphere used by Watt.
This higher pressure meant that a condenser and its associated pumps
were not needed. So small was Trevithick's engine that it could be
moved on a farm cart. He built a locomotive, which in 1804 made the
world's first locomotive haulage, pulling over ten tonnes of iron and
70 men in five trucks on a tramway between the Glamorgan Canal and
the Penydaren Ironworks.

George Stephenson (1781-1848), the son of the fireman at Wylam
colliery on Tyneside, had little formal education. However, he became the 'father of the railways' and the first president of the
Institution of Mechanical Engineers.

Stephenson's first locomotive, the *Blücher*, was built in 1814 while
he was working as the engineer for the Grand Allies, a syndicate of
collieries in the north-east of England. *Blücher*'s design was derived from other locomotives that Stephenson had seen working at local
collieries. There were, however, several original features including
flanged wheels, improved suspension, and a simpler transmission
system. Stephenson designed and built the Stockton and Darlington
Railway; it was opened in 1825 and was the first public railway in
the world. On this line he used rolled wrought iron rails, rather
than brittle cast iron rails; the first locomotive was the
Locomotion built by his son Robert (1803-59). On its inaugural run,
the *Locomotion* pulled 30 wagons which made up a load of about 100
tonnes.

Stephenson's next project was the Liverpool and Manchester Railway
which was planned to carry both goods and passengers. It was completed in 1830, despite considerable civil engineering problems. The
Liverpool and Manchester Railway had held the famous Rainhill trials
in 1829 to select the best locomotive type; these were won by Robert
Stephenson's locomotive *Rocket*. George Stephenson was in great demand to lend his name to the many railway schemes that had been prepared throughout Britain.

Robert Stephenson, like his father, was in great demand as an
engineering consultant. Among his major achievements were the London
and Birmingham Railway and the Britannia tubular railway bridge over
the Menai Straits.

Isambard Kingdom Brunel (1806-59) was not only a celebrated railway
engineer but also a steamship pioneer. He was the son of the French-
born British engineer Sir Marc Isambard Brunel (1769-1849) and worked
for his father in 1828 building the Thames tunnel between Wapping and
Rotherhithe. This tunnel, the first large-bore subaqueous tunnel in
the world, was successfully completed after 18 difficult years in

1843; it is now used by the London Underground Railway. To increase
the prosperity of Bristol, its merchants proposed a railway to link
Bristol with London. In 1833 Brunel was appointed engineer and
surveyor to the proposed Great Western Railway. The link took five-
and-a-half years to complete and was opened in 1841. The G.W.R. had
a gauge of 7 ft $0\frac{1}{4}$ in, not the 4 ft $8\frac{1}{2}$ in used by other companies and
copied by George Stephenson from colliery tramways. Brunel's broad-
gauge system was superior, providing for the first time express
trains that travelled at 60 miles per hour. However, the pressure
for standardisation was great since other railway companies had
adopted the narrower gauge. The building of future broad-gauge railways was banned by an Act of Parliament of 1846, although the broad-
gauge continued to be used in parts of the G.W.R. until the 1890s.
   During the nineteenth century and the early part of the twentieth
century the railway system in Britain grew in importance and British
engineers and capital constructed railways throughout the Empire and
in foreign lands. In the 1920s and 1930s, railways were under increasing competition from road transport and their profits declined.
They have now lost almost all of their freight traffic to the roads,
passenger traffic has dwindled, and many branch lines have been
closed. Steam locomotives were withdrawn in the 1960s and were replaced by diesel locomotives and increasing electrification.

1.5.2 Steamships
The earliest commercially successful steamship had its first voyage
on the Hudson River in 1807; it was designed by the American Robert
Fulton (1765-1815). Fulton's achievement was to design a hull that
would accommodate a steam-engine and yet operate profitably. The
engine that rotated the paddles was built by Boulton and Watt. The
ship carried up to 90 passengers at a speed of four knots and was of
100 tonnes capacity. She was originally christened the *Steamboat* but
after rebuilding was called the *Clermont*.
   Owing to the unreliability of their engines, steamships were provided with sails also. This practice continued until almost the end
of the nineteenth century. Sailing-ships reached their zenith in the
1850s and 1860s with clipper ships attaining speeds of 18 knots under
favourable wind conditions. Steamships, although slower, could steer
a straight course, did not depend on the wind, and could keep to a
regular timetable.
   The first Atlantic crossing partly under steam-power was made by
the American ship the *Savannah* in 1819. She was fitted with a
single-cylinder steam-engine having a 100 millimetre diameter cylinder and a stroke of just under two metres. In 1838 a British ship,
the *Sirius*, was the first to cross the Atlantic from east to west
mainly on her engines; the journey lasted 18 days.
   The Great Western Railway Company was interested in promoting
Bristol as a port. In 1835 they formed a subsidiary shipping company
and asked their engineer Brunel to build a steamship. This ship,
called the *Great Western*, made her maiden voyage to New York in 1838
arriving on the same day as the *Sirius* and completing the journey in
a record time of 15 days. The design of the *Great Western* was conventional; she had a wooden hull and was driven by paddle-wheels.
Her 750 horsepower steam-engines were built by Maudslay Sons and
Field.
   Brunel's second ship was the *Great Britain* which made her maiden
voyage in 1845. She had been planned as a paddle-steamer but was

built with a single screw, after the Swedish-born engineer John
Ericsson (1803-1889) had shown the screw propeller to be more efficient than paddle propulsion. The *Great Britain* was double the tonnage of the *Great Western* and when built was the largest ship in the
world. Her virtually indestructible iron hull was lying as a hulk in
the Falklands for many years before it was towed back to her home
port of Bristol where she has been restored as a museum.

The last of Brunel's three great ships was the *Great Eastern* which
made her maiden voyage in 1860. She greatly exceeded in size and
power any other ship; her wrought iron hull was 210 metres long and
her tonnage was eight times that of the *Great Britain*. It was not
until the *Celtic* was built in 1901 that the *Great Eastern's* tonnage
was exceeded. She was fitted with two single-expansion engines; one
drove the paddle-wheels, while the second drove the screw propeller.
The *Great Eastern* was a financial disaster; she was expensive to
operate and too large for the docks and shore facilities of the day.
Nevertheless she was a steady ship and had a hull design that incorporated many features used today. She is particularly remembered for
laying the first successful Atlantic telegraph cable in 1866. Worry
over the difficulties of building and launching the *Great Eastern*
were said to have contributed to Brunel's early death which occurred
just before the ship sailed on her maiden voyage.

Steel soon replaced wrought iron in shipbuilding and by 1891 80 per
cent of all ships being built were of steel. Double, triple, and
quadruple-expansion steam-engines with higher steam pressures were
used; these provided greater fuel economy and more space for cargo
and passengers. The Atlantic liners of the twentieth century adopted
the steam turbine to drive their multiple screw propellers. Speeds
increased to over 30 knots and lengths to over 300 metres. In the
1960s, air transport rapidly displaced the regular sailings of passenger liners throughout the world.

1.5.3 Road Transport

Road transport had to wait until a suitable engine was available to
drive vehicles. Although steam-engines had been used for road
vehicles it was the internal-combustion engine that enabled reliable,
economical and easily maintained road vehicles to be produced.

Steam Carriages

In 1769 a crude self-moving road engine was built by the French
military engineer Nicholas Joseph Cugnot (1725-1804). The first
practical steam carriage was developed in 1801 by Trevithick; two
years later one was running in the streets of London. Its inventor
however soon lost interest in this vehicle and turned his genius to
developing the steam railway locomotive. In 1832 a steam carriage
built by Goldsworthy Gurney (1793-1875) started running four times
each day over a distance of six miles between Cheltenham and
Gloucester. This horseless coach was faster and cheaper than the
horse-drawn stage-coaches with which it competed.

The development of steam-driven road vehicles was inhibited by
vicious legislation in 1831 and 1865. The legal requirement to have
a pedestrian carrying a red flag preceding a mechanically propelled
road vehicle was abolished in 1896. This led to a strong demand for
steam-driven traction-engines and goods wagons.

The construction of steam-driven road vehicles continued until the
1930s. By then, however, internal-combustion engines had been devel-

oped to such a high level of economy and convenience that steam-
engines could no longer compete.

The Internal-combustion Engine
Gas Engines

Gas Engines Gas engines were designed for stationary applications
but, as they were the precursors of petrol and diesel engines they
are included here for completeness. The first successful engine to
use town gas as a fuel was designed in 1860 by the Frenchman Etienne
Lenoir (1822-1900). Its design was based on that of a double-acting
horizontal steam-engine. A gas/air mixture was admitted to both
sides of the piston and fired electrically; an explosion on each side
of the piston was produced for each revolution of the crank. The
usual power take-off was a pulley attached to the crankshaft which
drove a belt. The running-cost was similar to that of a steam-engine
of equal power but the gas engine had a quick start-up and was more
convenient in use. Lenoir's engines were successful, although the
largest developed only 4 bhp.

In 1867 an improved gas engine was designed by the German Nikolaus
August Otto (1832-91). This engine ran at twice the speed and had
half the fuel consumption of Lenoir's engine. Five years earlier
Alphonse Beau de Rochas (1815-91) had published his proposals for a
four-stroke cycle; Otto used this cycle in his famous Otto 'Silent'
gas engine of 1876. The successful application of the four-stroke
cycle was an event of the greatest importance in the development of
the internal-combustion engine.

Owing to the patent protection on Otto's engine, a successful
British two-stroke gas engine was designed in 1880 by Sir Dugald
Clerk (1854-1932). In this engine the burnt gases were exhausted at
the end of the expansion stroke and the charge was admitted at the
start of the compression stroke. Prior to admission to the cylinder,
the charge was slightly pressurised by means of a pump. In 1891
Clerk introduced the idea of supercharging by raising the pump pressure, so packing more charge into the cylinder.

Fuels other than town gas were used for gas engines. Gas-producers
utilising cheap solid fuel were developed at the turn of the nineteenth century. Eventually, gas engines were superseded by diesel
engines and electric motors; many gas engines were converted to burn
liquid fuels.

Petrol and Diesel Engines A number of inventors tried to design an
internal-combustion engine that would burn liquid fuel. Early attempts failed because of the difficulty of metering, distributing, and
igniting the fuel.

An early success was that of Gottlieb Daimler (1834-1900) who, in
1884, designed a petrol engine that ran at 900 revolutions per minute.
One of these engines was fitted to a cycle in 1885, to create the
first motor cycle. Daimler's first engines used a wick carburettor
but in 1893 a more efficient Maybach jet carburettor was fitted.
Other inventors were working on petrol engines and new designs followed in quick succession.

The twentieth century has seen the sustained development of the
petrol engine. Efficiencies have greatly improved and reliability
has increased to the present almost trouble-free levels. The only
basic departure in petrol-engine design has been the Wankel engine.
Here reciprocating masses have been eliminated but fuel consumption
is not yet competitive with the conventional petrol engine.

The term 'diesel engine' is used to indicate a compression-ignition engine with airless fuel injection. The invention of this type of engine can be attributed to the British engineer Herbert Akroyd Stuart (1864-1937) who, between 1886 and 1890, took out a series of patents for a horizontal stationary oil engine. The fuel was sprayed 'solid' into a combustion chamber at the back of the working cylinder. The heat of compression and the residual heat in the combustion chamber ignited the charge. Before an Akroyd Stuart engine could be started its combustion chamber had to be pre-heated.

In 1892, Rudolph Diesel (1858-1913) took out a patent for a compression-ignition engine intended to use powdered coal as fuel. The engine, as eventually built in 1895, used oil, blasted into the cylinder by air, as a fuel; it had an efficiency greater than any contemporary heat engine. Although Diesel's engines were expensive to manufacture, they found early use in electricity generation and later use in marine applications. It was not until the high-speed diesel engine was introduced in the 1930s that the diesel engine became a rival to the steam-engine on the railways and to the petrol engine on the roads.

The Motor Car

The first reliable petrol vehicle was built in 1886 by Karl Friedrich Benz (1844-1929). This motor car had three bicycle-type wheels fitted with solid rubber tyres; the front wheel was steered by a tiller. The single-cylinder four-stroke engine had water-cooling and electric ignition; it could propel the car at 10 mph on the flat. An improved four-wheel version was introduced in 1890 and remained in production until 1902.

There was little public interest in the expensive, unreliable and uncomfortable early motor cars. In 1888 John Boyd Dunlop (1840-1921), a Scottish veterinary surgeon working in Ireland, invented the first pneumatic tyre with an outer casing and an inner tube. This did much to improve the comfort of the ride on both cars and bicycles. By 1900, cars had a maximum speed of 25 mph and could complete a cross-country journey at an average speed of about 12 mph. Considerable progress had been made by 1914. Engine speeds had risen to 3000 rpm; they could be controlled by the driver using an accelerator pedal which operated a butterfly valve in the carburettor. Electric ignition with variable timing had been introduced. It was no longer necessary to carry large amounts of water after the honeycomb radiator and water pump had been introduced. The early spoon-type brakes were replaced firstly by band brakes and then by expanding shoes operating against drums mounted on the rear wheels. The tiller was superseded by the steering-wheel which operated a worm and worm-wheel steering-mechanism; this arrangement improved steering accuracy and isolated the driver from road shocks. Engine performance improved considerably with power output increasing from about 5 to 20 bhp/litre. By 1914, sports cars could attain a speed of 85 mph. Passenger comfort was improved; first the windscreen and then side-doors were introduced and, finally, cars that totally enclosed their passengers were produced.

In 1908 the Model T Ford was first marketed; it did more than any other car to popularise motoring. Apart from its attractive price, this basic car was reliable and easy to drive. Mass-production techniques used by Henry Ford (1863-1947) were a major reason for the very low price of the Model T. By 1928, 15 million cars of this model had been produced. In addition to popularising motoring, Ford made a revolutionary contribution to manufacturing methods.

The 1920s was a period of design improvement. The cylinder block with its integral head and aluminium crankcase was replaced by a cast iron cylinder block to which the cylinder head was bolted. Distributor and coil ignition was introduced replacing the magneto. The multi-plate clutch gave way to the simpler single-plate clutch. Safety increased by the use of four wheel brakes and the wider use, towards the end of this period, of the all-steel body which had been pioneered by the Dodge Company in 1917.

The 1930s saw the introduction of independent front suspension, the syncromesh gearbox, and a more streamlined body shape. Since the Second World War there have been continual improvements in reliability, performance, safety, and comfort. An interesting development in the 1950s was the BMC mini designed by Sir Alex Issigonis (1906-). This car had a front-wheel drive and a transversely mounted engine, enabling 80 per cent of the total space to be provided for the passengers.

1.5.4 Flight
Balloons and Airships

The desire to fly has been with man from earliest times; it was first achieved in 1783 by Jean-François Pilâtre de Rozier in a hot air balloon constructed by the Montgolfier brothers. Later in the same year the French physicist Jacques Alexandre César Charles (1746-1823) accompanied by a friend reached a height of two miles and travelled a distance of 27 miles in a hydrogen-filled balloon.

Many attempts were made to steer balloons; the first real success came in 1852 with a powered dirigible balloon designed by the Frenchman Henri Giffard (1825-82). In Germany, Count Ferdinand von Zeppelin (1838-1917) constructed the first rigid airship in 1900. The development of airships continued until the 1930s when a series of major disasters terminated their general use.

Aeroplanes

The history of the aeroplane is confined to the twentieth century. The first powered flight of a heavier-than-air machine took place in 1903 in a biplane designed by two American bicycle mechanics, Wilbur Wright (1867-1912) and his brother Orville Wright (1871-1948).

Since then, development has been rapid for both military and civil use. Speeds have increased considerably from the 30 mph of 1903. The operational speed of airliners was about 100 mph in the 1920s, rising to 200 mph by the end of the 1930s then jumping from 350 to 500 mph with the introduction in 1952 of the first turbo-jet-propelled airliner, the De Havilland Comet.

Before 1940, the piston-type petrol engine was in almost universal use in aircraft, the cylinder arrangement being in-line or radial. Radial engines were usually air cooled, while water-cooling was used for in-line engines. At higher speeds and higher altitudes the propulsive limitations of a propeller driven by a piston engine became apparent. This led to the development of the gas turbine engine for aircraft.

Sir Frank Whittle (1907-) took out a patent on a turbo-jet engine in 1930 while he was a junior officer in the Royal Air Force. Little service or commercial interest was shown as it was considered that no materials could operate satisfactorily at the high temperatures generated in the turbine. Whittle eventually found financial backing and the Power Jets Company was formed in 1936. The first test flight of the engine took place in 1941 followed by its first operational use in the Gloster Meteor fighter at the end of the Second World War. At about the same time as Whittle's engine came into service, the Germans fitted a jet engine of their own design to a Messerschmitt 262 fighter aircraft.

Since the 1950s, the gas turbine engine has become the principal method of aircraft propulsion. It has approximately twice the power/ weight ratio of a piston engine and each engine can be designed to develop considerably more power. The most powerful piston-type air- craft engine was the 28-cylinder Pratt and Whitney Wasp Major radial engine which developed 3500 hp.

Helicopters
The first helicopter flight was in 1907 when the Frenchman Paul Cornu and a passenger were lifted for a few minutes in a machine construct- ed by Cornu. The first commercially successful helicopter was de- signed by the Russian-born American Igor Ivanovich Sikorsky (1889- 1972). It first flew in 1939 and has been rapidly adopted for mili- tary and specialised civil use.

1.6 ELECTRICITY

1.6.1 The Pioneers
Although the presence of electricity had been observed in nature from earliest times, it was not until the nineteenth century that a start was made to generate and to use electricity. This section outlines some of the early discoveries; subsequent sections give an equally brief treatment to telecommunications, electrical power, and the transistor. In 1745, Pieter van Musschenbroek (1692-1761) of Leyden discovered a method of holding a charge of static electricity in a glass jar which became known as a Leyden jar. The American Benjamin Franklin (1706-90) was a pioneering investigator; he is particularly remembered for his kite experiment which led to the invention of the lightning conductor. He proposed in 1747 the one-fluid theory, the first reasonable theory of electricity. The French military engineer and physicist Charles Augustin de Coloumb (1736-1806) made the next major contribution when he produced the first quantitative law of electricity; this was that the force of attraction and repulsion between small charged spheres is inversely proportional to the square of the distance between them. Alessandro Volta (1745-1827), an Italian professor of physics, invented in 1799 the first electric battery. This consisted of alternate discs of silver and zinc sepa- rated by absorbent material soaked with water. Volta's important invention provided a source of current electricity for experimental work by other investigators.

In 1820 the Dane, Hans Christian Oersted (1777-1851) discovered electromagnetism. Later in the same year, André-Marie Ampère (1775- 1836) reported to the French Academy the magnetic effect of current flowing through coils of wire. In the following year the British chemist and physicist Michael Faraday (1791-1867) showed that a wire carrying current would continuously revolve if subjected to a mag- netic field, thereby paving the way for the invention of the electric motor. Subsequent experimental work by Faraday established the basic principles of the transformer and of the mechanical generation of electricity. In 1826 the German physicist Georg Simon Ohm (1787- 1854) proposed his well-known law that linked the current, resistance, and electrical potential of a circuit.

James Clerk Maxwell (1831-79), the brilliant Scottish-born physicist, did much to interpret mathematically Faraday's work on the electromagnetic field; in 1865 he published his celebrated paper 'A Dynamical Theory of the Electromagnetic Field'. An important prediction made from Maxwell's work was that alternating electromagnetic disturbances produced electromagnetic waves that travel with the speed of light. It was the German physicist Heinrich Rudolf Hertz (1857-94) who, in 1887, produced the experimental evidence that confirmed Maxwell's predictions on electromagnetic waves and laid the foundations for future work that culminated in radio and television.

1.6.2 Telecommunications
The Electric Telegraph
Ampère had suggested in 1820 that messages could be transmitted by having an electrical circuit for each letter with a magnetic needle at the receiving end. This idea was taken up by a number of inventors. The first practical system came from the work of William Fothergill Cooke (1806-79) and Charles Wheatstone (1802-75) who built a one-mile long telegraph line in London between Euston Station and Camden Town. The first message was sent along this line in 1837. Cooke and Wheatstone's early system required up to six wires for message transmission. In the same year of 1837 the American Samuel Finley Breese Morse (1791-1872) sent his first message along one-third of a mile of wire. Shortly afterwards he introduced the Morse code; this audible dot-and-dash code greatly facilitated message transmission. Telegraph systems were rapidly expanded in the 1840s. The first major submarine cable was laid between Dover and Calais in 1850. The first transatlantic cable was completed in 1858 but failed after only three months of use. A second and successful cable was laid between Ireland and Newfoundland by the *Great Eastern* in 1866. Although telegraph traffic grew steadily in the nineteenth century, it has declined in the present century because of competition from the telephone.

The Telephone
The Scottish-born Alexander Graham Bell (1847-1922) is credited with the invention of the telephone by virtue of a patent filed in 1876. Another inventor, Elisha Gray, filed a patent for a similar type of instrument a few hours afterwards.

Bell emigrated to Canada in 1870 and went on to the United States in 1872. By 1877, the telephone was in commercial use in America and thereafter its use rapidly spread throughout the world.

Radio
The advantage of sending point-to-point messages by means of electromagnetic waves was that no wire or cable connection was needed. An early application was wireless telegraphy, where Morse code messages were used for ship-to-ship and ship-to-shore communication. It was not until the 1920s that the use of radio waves for the transmission of the voice and music was exploited. In 1896, an Italian, Guglielmo Marconi (1874-1937), transmitted and received a coded wireless message over a distance of almost two miles. The first transatlantic message was sent by Marconi from Cornwall to Newfoundland in 1901. The British physicist Sir Oliver Lodge (1851-1940) introduced his selective tuner in 1898, a device that greatly improved the detection of radio signals. The first wireless valve, a diode, was invented in

1904 by the British physicist Sir John Fleming (1849-1945). In 1906,
Lee de Forest (1873-1961), an American inventor, inserted a metal
grid between anode and cathode. De Forest's triode provided much
better control of current flow than was possible with the diode.
Four years later in 1910 the voice of Enrico Caruso, the famous
Italian opera singer, was broadcast.

By 1920, efficient radio transmitters and receivers were being
manufactured. In 1922, the British Broadcasting Corporation was
formed. Reception had been greatly improved by the invention of the
feedback circuit in 1914 and the superheterodyne circuit in 1918.

Television

The first limited television programme was shown in 1929 by the
British Broadcasting Corporation. The BBC used a mechanical scanning
system developed by John Logie Baird (1888-1946). Although the idea
of electronic scanning had been proposed as early as 1908, it was not
until 1929 that Vladimir Kosma Zworykin (1889-   ) a Russian-born
American, demonstrated his electronic scanning system. This proved
to be greatly superior to the mechanical system which it rapidly replaced.

1.6.3 Electrical Power

Electricity is the most convenient form in which energy can be supplied for industrial and domestic use. The ease with which it can be
transmitted over long distances has made it widely available and
industry is no longer restricted to areas where water-power or cheap
coal is available.

Generation

It is interesting to note that until about 1870 the storage battery
was the principal source of electrical power. A number of inventors
contributed to the development of electrical generators; perhaps the
most notable was the Belgian-born Zénobe Théophile Gramme (1826-1901).
During the 1870s and 1880s, a large number of Gramme generators,
driven by steam-engines, were installed.

Distribution

In the early days, electricity was largely used for lighting purposes,
firstly to supply carbon arc lamps, available in an efficient form in
1876, and later for incandescent lamps, developed for domestic use by
the American, Thomas Alva Edison (1847-1931). The original distribution systems were supplied by direct-current generators with voltages of between 100 and 200 volts. This arrangement was inefficient
owing to the large power losses in transmission. In 1881, an alternating-current electrical system was demonstrated in London. Alternating current had the advantage that transmission losses could be
minimised since power could be moved over long distances at high
voltages but at low currents. The high-voltage bulk supply could
then be converted into a low voltage for local distribution by means
of transformers. Important early work on alternating-current systems
was done by the Croatian-born American Nikola Tesla (1856-1943).
While still in Europe, he conceived the idea of polyphase alternating
current and the induction motor. He emigrated to the United States
in 1884 and set up his own company producing alternating-current power
systems; this he sold to the Westinghouse Company in 1899.

Lighting

There have been considerable improvements to the efficiency of electric lamps. Edison's original carbon-filament lamps had a light output of 1.7 lumens per watt; this was improved to 4.2 lumens per watt
by the metallised carbon-filament lamp of 1905. Tungsten-filament
lamps were introduced in 1902 and largely superseded carbon-filament
lamps within 15 years. Modern tungsten-filament lamps produce 22
lumens per watt. Fluorescent lamps, which were first introduced in
Germany in the early 1930s, produce 65 lumens per watt.

1.6.4 The Transistor

This important development was the result of team-work at the Bell
Telephone Laboratories. Those concerned were the Americans William
Shockley (1910-   ), John Bardeen (1908-   ), and Walter Houser Brattain
(1902-   ). The public announcement of their discovery in 1948 received little publicity but, when production difficulties had been
overcome, transistors began to replace the thermionic valve. The
transistor has many advantages over the valve; they include its very
small size, long life, and low power requirements.

1.7 THE COMPUTER

Computers have an important influence on the work of engineers. The
design engineer can use them to obtain rapid and accurate answers to
his design calculations. He also can use computer-aided design (CAD)
to make the designing and drawing processes more productive. The production engineer can employ computers to control processes, machines,
and sections of the factory. In addition, computers can store, mani-
pulate, and produce data for management information systems.

The design/manufacturing interface has now been bridged by computer-
aided engineering (CAE) which enables designs to be directly trans-
lated into instructions for computer-controlled machines to manufacture
the parts.

1.7.1 Early Computers

The British mathematician Charles Babbage (1791-1871) was not the
first person to design a calculating machine; however, his 'difference engine' was important as its design incorporated almost all the
essential features of the digital computer. The machine was mechanically operated and extremely difficult to manufacture using the
equipment of the day. Despite 13 years of work and a large expenditure of money, Babbage abandoned the project in 1833 with an unfin-
ished machine but with the principles of the digital computer established.

Although the slide rule has prior claim, many consider that the
first analogue computer was made in 1876 by the Ulster-born mathematician and engineer Lord Kelvin (1824-1907). This, like Babbage's
difference engine, was a mechanically operated computer. Kelvin
designed it to predict tide times.

Developments up to about 1940 included a printing calculator, made
in 1854 by the Swedish engineer Georg Scheutz (1785-1873), and,
towards the end of the nineteenth century, the introduction of
accounting machines and the use of punched cards in calculating
machines.

The idea of using punched cards to carry information was first used
around 1800 by the Frenchman Joseph Marie Jacquard (1752-1834). He
used punched cards in the form of an endless belt to control patterns

produced by his weaving looms. The German-born American Hermann
Hollerith (1860-1929) used separate rectangular punched cards to
carry information. Hollerith employed electrical sensing devices to
enable the hole positions on the cards to be sensed so that the
information could be sorted and analysed. An early use of punched
cards was in the compilation of the United States census of 1890.
Over the years, punched-card equipment and systems were improved;
however the excessive time required to perform simple arithmetic
operations, such as multiplication, limited their usefulness.

1.7.2 The Modern Computer

Work started on the first electronic computer in 1942 at Pennsylvania
University. The machine came into use in 1946 and was called the
ENIAC (Electronic Numerical Integrator and Calculator). The ENIAC
used 18,000 thermionic valves and occupied a floor area of 200 square
metres. It had a very small memory of 20 words consisting of ten
figures each and was restricted to a program of not more than 300
words. A further step forward was taken in 1950 with the production
of the UNIVAC I (Universal Automatic Computer). This computer was
designed to incorporate the proposals of John von Neumann (1903-57)
the Hungarian-born American mathematician. Modern computers can be
said to be descended from the UNIVAC I.

The second generation of electronic computers came at the end of
the 1950s when valves were replaced by transistors; this reduced size
and improved reliability. Methods of data storage were greatly
improved and enlarged. Rapid random access to data storage using
discs rather than tapes became available and high-speed printers that
more nearly matched computation speeds were developed.

Finally, the third generation of computers came when the first
mass-produced integrated circuit became available around 1964. These
circuits have become progressively smaller and less expensive.

The availability of the integrated circuit gave birth to the micro-
processor. In 1969 the American company Intel produced a very
elementary central processing unit (the electronic logic circuit
constituting the brain of the computer) on a single chip. It proved
to be too slow for its intended application but Intel decided not to
abandon the project and their chip found a market as a programmable
logic device. Microprocessors appeared in pocket calculators in the
early 1970s; they have since been widely used as inexpensive control
devices in a wide range of industrial and domestic equipment.

1.8 THE ENGINEERING INSTITUTIONS

1.8.1 The Beginnings

In 1662 Charles II granted a royal charter to the Royal Society which
had been founded two years earlier to promote 'Physico-Mathematicall
Experimentall Learning'. The Royal Society provided a forum for
science and, to some extent, engineering.

Unlike France, where engineering tended to be centrally organised
and encouraged, British engineering grew from the inventive genius
of individuals. In the eighteenth and nineteenth centuries, British
inventors formed groups where they discussed their work. The Society
of Civil Engineers was founded in 1771 by John Smeaton (1724-92).
Smeaton was the first Englishman to call himself a civil engineer;
this was to distinguish himself from those who practised military
engineering.

Another of these early societies was the Lunar Society which met
in Birmingham on the Monday nearest to the full moon. This society
included the distinguished membership of Matthew Boulton, James Watt,
John Smeaton, Joseph Priestley the scientist, and Josiah Wedgwood the
potter.

1.8.2 The Institution of Civil Engineers

In 1818 a small group, including William Maudsley and his partner
Joshua Field, met at the King's Tavern, Cheapside to set up a body
to rival the Society of Civil Engineers. H.R. Palmer, who brought
together this founding group, said: 'An engineer is a mediator between the philosopher and the working mechanic and like an interpreter between two foreigners must understand the language of both...
Hence the absolute necessity of possessing both practical and
theoretical knowledge'. That statement is as relevant today as when
it was made.

Unlike earlier associations of engineers, the Institution of Civil
Engineers had come to stay. It acquired its own premises in London
firstly in Buckingham Street and then at Cannon Row. Today its
headquarters are in Great George Street next to those of the
Institution of Mechanical Engineers in Birdcage Walk.

1.8.3 The Institution of Mechanical Engineers

This institution grew out of the reluctance of the canal engineers,
who in the 1840s were influential in the Institution of Civil
Engineers, to accept railway engineers into their ranks. The
Institution of Mechanical Engineers is said to have sprung from the
conversation of a group of railway engineers who were sheltering
from the rain in a platelayers' hut at an engine trial in 1846. They
were discussing the affront to George Stephenson who had been asked
to submit an essay on his capacity as an engineer before he could be
admitted to the Institution of Civil Engineers. Soon afterwards, a
meeting of railway and manufacturing engineers was convened where it
was resolved to form an Institution of Mechanical Engineers. The
Institution was formally founded in 1847, with George Stephenson as
its first president. It now has more corporate members than any
other British engineering institution and incorporates the previously
independent Institution of Automobile Engineers and Institution of
Locomotive Engineers.

1.8.4 The Institution of Electrical Engineers

This institution was founded in 1871 as the Society of Telegraph
Engineers with Sir William Siemens as its first president. It
received encouragement from the Institution of Civil Engineers and
held its meetings for some years at their headquarters. The title
was changed to the Institution of Electrical Engineers in 1888.

1.8.5 The Institution of Production Engineers

Manufacture and management is the chief concern of this institution
which was founded in 1921. In recent years the Institution of
Mechanical Engineers has become more interested in production engineering; hence there is a considerable overlap of interest.

1.8.6 The Engineering Council

In 1965 the 13 chartered engineering institutions formed the Council
of Engineering Institutions with the aim of having a single body to

represent the different branches of the profession. In 1981 the
Engineering Council was formed to take over and to strengthen the work
of the Council of Engineering Institutions.

1.9   ENGINEERING EDUCATION

In contrast with France and Germany, engineering education in Britain
in the eighteenth century and much of the nineteenth century was
inadequate. Towards the end of the eighteenth century, France led
the world in the application of scientific knowledge to engineering
but Britain rapidly outpaced France in engineering achievement. As
a result, the myth was created that British engineers possessed a
talent and inspiration that did not need a scientific basis. This
illusion of innate superiority persisted well into the nineteenth
century. The almost total lack of scientific training for most
engineers until the second half of the nineteenth century is considered by many to be the prime reason for Britain's loss of engineering leadership after 1850. The Institution of Civil Engineers did
not introduce entrance examinations until 1897; the first entrance
examinations for the Institution of Mechanical Engineers were in
1912.
   A large proportion of nineteenth century engineers were trained by
pupillage by which parents paid large sums of money to individual
principal engineers for their sons to be trained by them. During the
eighteenth century it was not always necessary that a man should
serve an apprenticeship before he could practise a craft and in 1814
the last legal restriction was lifted. Apprenticeships, however, did
not cease and many companies continued to offer excellent craft
apprenticeship.

1.9.1   The Universities
The universities of Cambridge and Glasgow have had a long-established
interest in engineering science and the work of Glasgow was outstanding in the nineteenth century. London University was early in introducing engineering education; King's College, London set up a course
in civil and mechanical engineering in 1838 and University College
founded three engineering chairs during the period 1841 to 1846. A
number of private schools were also founded at about the same time to
train engineers. The civic universities set up between 1871 and 1880
at Newcastle, Manchester, Leeds, Sheffield, and Liverpool and subsequent foundations, notably Birmingham, have made substantial contributions to engineering education. One of the leading engineering
colleges, The Imperial College of Science and Technology, was a comparatively late foundation, being an amalgamation in 1907 of the
Royal School of Mines, the City and Guilds of London Technical
Institute, and the Royal College of Science. In 1980, 48 universities
and university colleges offered degree courses in engineering.

1.9.2   Technical Colleges and Polytechnics
There were a number of attempts in the first half of the nineteenth
century to educate factory workers by means of lectures; the best
known of these was the Mechanics' Institute movement. Owing to a
general absence of elementary education, these attempts were largely
unsuccessful.
   In 1889 the newly formed County and County Borough Councils were
granted powers to provide technical education. To finance this

expenditure they were given part of a tax on whisky and, if they
wished, they could levy a one-penny rate. A large number of technical colleges were soon built, the 'whisky money' being the main
stimulant. A system of national examinations appropriate to the
evening classes held in these colleges was provided by the City and
Guilds of London Institute, founded in 1878.

Soon after the First World War, the National Certificate Awards
were instituted in various branches of engineering. An ordinary
certificate was awarded after three years of part-time study and a
higher certificate after two additional years. Holders of appropriate Higher National Certificates obtained prior to 1970 were
academically qualified for corporate membership of the major engineering institutions. Today an appropriate engineering degree is
normally required of those seeking corporate membership of the
engineering institutions.

National Certificate examinations have now been replaced by the
examinations of the Business and Technician Education Council which
has also absorbed the work at technician level of the City and Guilds
of London Institute.

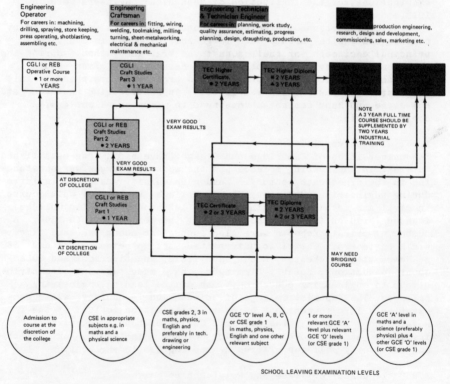

Figure 1.8   A guide to engineering education (by courtesy of the
Engineering Careers Information Service)

In 1970 the Government formed the polytechnics from a number of
leading technical colleges usually in amalgamation with a college of
art. There are now 30 polytechnics offering high-level engineering

courses. In 1979, 28 polytechnics and 12 other institutions operated engineering degree courses validated by the Council for National Academic Awards. In the same year about one-third of all entrants to first-degree courses in engineering were admitted to these non-university institutions.

Engineering education now falls into four levels: operator, craftsman, technician, and professional with transfer between levels. A guide issued by the Engineering Careers Information Service is shown in figure 1.8.

1.10 THE IMPACT OF ENGINEERING TECHNOLOGY

There is no doubt that the development of engineering technology in the nineteenth and twentieth centuries has transformed human existence throughout most of the world.

In particular, muscular effort and drudgery have been taken out of work both in the factory and the home. An average person can sustain a muscular effort equivalent to 35 watts during the length of the working day. Today the operator of a relatively small machine tool has under his control the muscular effort of several dozens of men, while the driver of a car can command an effort equal to that of a thousand men. As well as supplying adequate power, engineering technology ensures that power is supplied in an appropriate form. For instance, the equivalent of the muscular effort of a million people enables the pilot of an airliner to convey hundreds of passengers for thousands of miles at 500 miles per hour.

Engineering has supplied not only the power but the means of controlling it automatically. An early development was Watt's centrifugal governor used to control the speed of his rotative steam engine. In the 1920s the first reliable and fast-acting servo-mechanisms became available. The industrial application of automatic control has been rapid in the latter part of the twentieth century. The operation of oil refineries and many chemical plants is virtually automatic and sections of factories operate automatically. A parallel development, starting in the 1950s, has been the automation of much routine clerical work by using computers and word processors.

These changes have enabled output per employee to be increased to meet the world's explosive demand for goods and services while allowing the working week to fall from 60 hours in the nineteenth century to 40 hours or less today.

Transport developments in the last 150 years have shrunk the size of the world and have brought international and intercontinental travel within the means of the average wage-earner. Television enables events throughout the world to be seen as they occur and telephone subscribers can now dial direct to many parts of the world.

The undoubted benefits brought by engineering technology, however, have not been distributed evenly and most of the population in the world's poorer countries still live in dire poverty.

Although there are few who would wish to return to the hard, insanitary, disease-ridden conditions of the early nineteenth century, not all is on the credit side. There is industrial pollution and the availability of nuclear weapons, with their well-engineered methods of delivery, has provided civilisation with the means of its own destruction. Let us hope that no nation will be foolish enough to return the world to the Dark Ages.

FURTHER READING

Armytage, W.H.G., *A Social History of Engineering* (Faber, London, 1976).

Finniston, Sir M. (Chairman), *Engineering our Future*, Report of Committee of Inquiry into the Engineering Profession (HMSO, London, 1980).

Greaves, W.F. and Carpenter, J.H., *A Short History of Mechanical Engineering* (Longmans, Harlow, 1978).

Rolt, L.T.C., *Victorian Engineering* (Penguin Books, London, 1974).

Woodbury, R.S., *Studies in the History of Machine Tools* (MIT Press, Cambridge, Mass., 1973).

2   Industrial Organisations and their Environment

In this chapter the various types of organisation for which an
engineer is likely to work are discussed. The second part of the
chapter considers some of the economic and social pressures acting
on an industrial enterprise.

2.1   PRIVATE SECTOR ORGANISATIONS

2.1.1 Historic Developments
The English chartered companies have had a long history. The famous
East India Company was formed in 1600, followed later in the seventeenth century by the Hudson's Bay Company and the Bank of England.
   During the eighteenth century almost all trading was carried out by
sole traders or partnerships. As a result, the capital available to
launch business ventures was limited; it had to come from individuals
closely connected with the venture rather than from the public at
large. This difficulty was overcome for certain types of enterprise
by founding companies by a private Act of Parliament. These were
called statutory companies and the subscription of capital was open
to the public; many gas, railway, and electricity companies were
established in this way.
   A major step forward came with a series of Companies Acts. The
first, in 1844, provided a simple way of forming companies to which
any member of the public could subscribe. In 1855 the Limited
Liability Act was passed. This Act limited the liability of shareholders to the loss of the money they had paid for their shares; no
longer did shareholders have to pay the debts of a failed company out
of their own pockets. Since 1855 various Companies Acts have been
passed; the present legislation is contained in the Companies Acts of
1948, 1967, 1976, and 1980.

2.1.2 Sole Traders
Persons running a business on their own are sole traders. They are
solely responsible for the success or failure of the business and
have no special legal position. The profit earned is their income.
Any losses have to be made good from their personal resources. If
money is borrowed from the bank they may be asked to deposit the
deeds of their house as security against failing to repay the loan.
   If the business trades in a name other than that of its proprietor
his name must be displayed on the premises and on the letterheads he
uses.

2.1.3 Partnerships
Partnerships are formed to increase the capital of a business, to
bring in special expertise, and to share the risks and management of
the firm. Most partnerships provide professional services; in cer-

tain professions such as medicine, dentistry, the law, accountancy, and the Stock Exchange, partnerships are the normal form of trading. Engineers working as consultants may form partnerships although most consultancies are limited companies.

The law concerning partnerships is contained in the Partnership Act of 1890 and under the Limited Partnership Act of 1907. Partners can draw up their own partnership agreement thereby varying certain provisions of the Acts. Partners are regarded as individuals by the law and are personally liable for the debts of the partnership except in limited partnerships.

Limited partnerships have one or more general partners who remain personally liable for the payment of any debts. The limited partners lose their limited liability if they take part in the management of the firm.

As with sole traders, partnerships have to display partners' names at the business premises and on their letterheads.

2.1.4 Companies

Companies account for about 15 per cent of all businesses in Britain and are the backbone of our industrial society. They have a separate legal identity with legal rights and duties.

Liability for Debt

Unlike sole traders and partnerships the shareholders in most companies are not personally liable for the debts of the company. If the company has limited liability, the shareholders can lose the money they have invested but no more. Many newly issued shares start life in partly paid form and anyone purchasing them has to pay the balance due within a relatively short period. Although almost all companies are limited by share, a few are limited by guarantee or have unlimited liability. Companies that are limited by guarantee can call upon their shareholders to pay whatever sum of money each has previously undertaken to pay.

Private and Public Companies

Of the 600,000 registered companies in Britain, the vast majority are private companies. Private companies are typically small to medium-sized family businesses frequently starting life as sole traders or partnerships. In a private company the number of shareholders varies between a minimum of two and a maximum of 50; there are restrictions on the transfer of shares and the public must not be invited to subscribe capital.

Public companies can have an unlimited number of shareholders, shares may be freely transferred, and the public can be invited to purchase shares. Most of the larger public companies have the price of their shares quoted on the Stock Exchange. This provides a very free market for the shares and greatly facilitates the raising of additional capital. The Stock Exchange has stringent requirements that must be met before it will allow the shares of a company to be bought and sold there. About a quarter of the public companies have Stock Exchange quotations.

Multi-national Companies

These companies have their headquarters in one country but also perform production, marketing, finance, and personnel functions in other countries. There are many American and European multi-national companies; they are particularly strong in the oil, vehicle, and electrical industries. An example of a British multi-national is the Rio-Tinto Zinc Corporation (RTZ); this company has interests in almost every major metal and fuel. In 1981, RTZ had an annual turnover of £3000 million; it operated through 37 principal subsidiary and associated companies incorporated in 11 different countries.

Associated and subsidiary companies are subject to the laws of the countries in which they are incorporated. In addition they are also subject to special laws and regulations that prevent loss of tax if goods are transferred between countries at prices designed to reduce the overall tax liability of the parent company.

Multi-nationals have the advantage of easy access to capital, management strength, the ability to switch production to low cost areas, and considerable influence over the government of many small less-developed countries. Difficulties can arise when the policies of a multi-national company come into conflict with those of a country in which it is operating. Also, resentment can occur when product research, development, and design are confined to the country in which the parent company is registered. However, most multi-nationals are conscious of their obligations to host-countries and create prosperity and employment beyond that which would be possible without their presence and expertise.

2.1.5 Formation of a Company

It is a relatively simple matter to form a company by lodging the following documents with the Registrar of Companies.

(1) *Memorandum of Association* This states the name of the company and the objects that it intends to pursue; it includes a statement of how the capital is organised.

(2) *Articles of Association* This document sets out the internal rules by which the company is to be organised. It provides information on such matters as the calling of shareholders' meetings, the voting rights of various types of shareholders, and the powers of directors. A special resolution of the shareholders can alter the Articles of Association.

(3) A declaration that the company will comply with the requirements of the Companies Acts.

(4) A statement of the nominal value of the share capital and how it is to be divided up into shares.

Notice of the address of the registered office must be lodged within 14 days of the incorporation of the company and particulars of the directors and secretary must be given within 14 days of the appointment of the first director.

2.1.6 Other Types of Private Sector Organisation

Banks and insurance companies have their activities controlled by special regulations in addition to having to comply with the Companies Acts. Co-operative societies form a distinct group of business organisation and are registered under the Industrial and Provident Societies Act. A workers' co-operative can be registered either as a co-operative society or as a company.

2.2 PUBLIC BODIES

A considerable number of engineers work in the public sector as civil servants or as employees of public corporations and local authorities.

2.2.1 Public Corporations

The National Coal Board, British Rail, and the British Steel Corporation are examples of public corporations engaged in industrial
activities. Most public corporations are set up by an Act of Parliament and are answerable to an appropriate government Minister. Each
corporation has a board which is a corporate body with a legal
identity. The board is appointed by the Minister and functions in
a way similar to the board of directors of a commercial company,
taking decisions on such things as pricing and capital expenditure.
Sometimes board decisions are overruled for political reasons, particularly in the field of price fixing. There are no shareholders to
satisfy but the activities of public corporations can be debated in
Parliament and are subject to investigation by Select Committees of
the House of Commons.

It is possible for governments to control the activities of industrial companies, either by outright purchase or by acquiring a controlling interest. The type of company acquired by government is
typically a large organisation in financial difficulties which, in the
national interest, should not be allowed to go into liquidation; examples are Rolls-Royce and British Leyland.

There are a number of semi-independent bodies established by the
government as instruments of administration and control; these include the National Enterprise Board and the Regional Health Authorities.

2.2.2 Local Government

The responsibility for managing local authorities is shared between
elected councillors and permanent officials; the officials advise the
councillors and implement council decisions. The council, which is
the governing body, sets up committees and sub-committees in which
much of its business is done.

Central government has considerable control over local authorities
through the Department of the Environment. Should there be any
maladministration, the Minister can vary the amount of the Rate
Support Grant. The accounts of local authorities are subject to
audit by auditors appointed by the Department of the Environment.

2.3 USE OF RESOURCES

A problem facing all national economies is to satisfy the demands of
its consumers for goods and services. The resources to do this, that
is, labour, material, land and capital, are limited and in consequence
there is a problem of how these should be used. There are two basic
approaches.

Firstly there is the centrally planned system, as used in the USSR
and Eastern Europe, where the state decides what should be produced.
This system can operate satisfactorily if the demand is unsophis-
ticated and the population politically quiescent.

Secondly there is the market system which has greater flexibility;
here society determines through its purchases how resources are to be
used. The economies of the richer non-communist countries are pre-
dominantly market orientated.

2.3.1 A Model of the Market System

Although the assumptions made in this model ignore many realities, it
is sufficiently relevant to the British economy to be considered.

The first assumption that is made is that people are self-seeking;

employees try to increase their income, producers try to make more profit, and consumers attempt to satisfy their insatiable demand for goods and services. It is also assumed that there is free competition with producers competing against each other for sales to consumers and for the resources needed to manufacture their products.

In this situation, demand and price will be linked. When demand for a product exceeds supply, the price of the product will rise and producers will be encouraged to increase output. If, however, supply exceeds demand, prices will fall and producers will reduce output. Therefore in the market system it is the consumers' preferences that determine output and hence the use of resources.

Because the consumer is purchasing in a competitive market and attempting to satisfy his needs from a limited income, he will seek those products that offer him the best value. This search for value will benefit the producer having the best design as well as the most efficient production and distribution system. The competition that exists in the market will prevent producers from making excessive profits. When demand exceeds supply, the ensuing price rise will cause existing producers to increase their output and induce new producers to enter the market. The result will be that supply catches up with demand and prices will stabilise. If supply overshoots demand, the price will fall.

It should be appreciated that the validity of this model will be affected by the imperfect assumptions made in its construction. The assumption of perfect competition in the market is unlikely to be correct. Most producers will attempt to influence consumer preferences by advertising rather than allow consumers to make a free choice. Demand will not instantaneously respond to price changes; there is bound to be some time-lag. In fact, greater demand for a product may not, as expected, cause producers to increase their prices since they may anticipate eventual cost reductions resulting from a greater volume of production.

Lastly, even in a market economy, there is likely to be some measure of central planning which will further distort the operation of the model.

2.4 OBJECTIVES OF THE FIRM

The traditional assumption that profit maximisation is the prime objective of the firm is the subject of considerable discussion and some dispute. Space does not permit a consideration of the various models that have been proposed to explain how companies select their objectives. In this section we will discuss the major objectives of profit, growth maximisation, and corporate survival, together with the social objectives of the firm.

Profit growth is likely to be the prime concern in most companies with other objectives being subservient to it. However, an acceptable achievement in competing objectives will also be required. The relative importance given to the various objectives will be determined by the board of directors who will be influenced by the pressures acting on them.

2.4.1 Profit Maximisation

It has been traditionally assumed that profit maximisation is the overriding objective of the firm. This can be accepted as a reasonable assumption with sole traders, partnerships, and small companies, for here the life-style and social standing of the decision-makers are directly affected by the profitability of their enterprise.

In large companies the directors and senior managers tend not to be the owners of the business and therefore are less likely to be concerned with profit maximisation. A typical large quoted company will be owned by tens of thousands of private individuals and a smaller number of institutions such as insurance companies, investment trusts, and pension funds. Individual shareholders often have little commercial knowledge and are unlikely to form pressure groups. Institutional shareholders often possess considerable financial knowledge and power but are reluctant to criticise openly the conduct of the companies in which they have invested. Although directors can be removed from office by shareholders who collectively own more than 50 per cent of the voting shares, the voting-out of directors is an unusual event.

However, the low risk of removal from office by shareholders does not mean that directors of large companies can be indifferent to profitability. Firstly, if a company fails to make a profit over a prolonged period, it is likely to go into liquidation. Further, poor profit performance will cause the price of the company's ordinary shares to fall, thereby enabling its assets to be acquired cheaply by another organisation making a take-over bid. When a company is taken over, its directors and senior managers are often displaced.

Apart from the penalties for low profitability, high profitability has considerable advantages. A company can afford to spend adequately on marketing, on product development, and on manufacturing equipment, so keeping itself in a strong competitive position. It can grow either by ploughing money back into the business or by the take-over of other companies, being assisted in this by its own high share price.

2.4.2 Growth Maximisation

Growth is likely to be a major objective of a company as there can be significant reductions in product cost when size increases. These economies are referred to as 'economies of scale' and result from the following factors.

(1) More efficient production methods are possible with higher-volume manufacture.

(2) Larger orders give greater bargaining strength with suppliers of materials, parts, and services and result in cheaper purchases.

(3) Promotion and distribution costs per unit sold are lower.

(4) Additional finance is more easily available through the issue of shares or loan stock, or by overdraft.

(5) Better and more highly specialised management ability can be engaged.

(6) In process industries and in bulk transport considerable saving in plant cost is possible. Cost tends to be proportional to surface area (ℓ^2), whereas capacity is proportional to volume (ℓ^3). This is the argument for the construction of supertankers and large chemical plants.

There can be diseconomies of scale; large organisations are difficult to manage and motivate and may become riddled with inefficiency. However, with vigorous management and adequate decentralisation, the balance of advantage usually lies with the large organisation.

2.4.3 Social Objectives

Despite the ultimate duty of a company being to its shareholders, it also owes a duty to its employees, its customers, its suppliers, and to the public at large. Society will allow companies to exist and expand so long as they provide the social and economic benefits expected of them. These expectations rise as people grow to anticipate more from life. Three levels of company social responsibility can be distinguished.

At the primary level a company does no more than meet its basic obligations. Most of these obligations must be met otherwise the company risks legal penalties.

The second level is reached when a company voluntarily accepts more than its minimum obligations. The additional expenditure at this stage is likely to be worth while. For example, the provision of better-than-average terms and conditions of employment can enhance the company's reputation and attract staff who are significantly better than those employed by most of its competitors.

The third level of social responsibility is one of much wider social awareness in which the company considers its responsibility towards maintaining the framework of the society in which it operates. At this level the company looks outwards as well as inwards and severe conflicts can arise between the interests of shareholders, employees, and the public. Here there is almost certain to be some loss of profit.

Many companies are keenly aware of their social responsibilities and sponsor projects of environmental, cultural, and sporting interest. Some advertising advantage is gained but this is often small compared with the cost.

Specific examples of some of the questions that a company may have to answer when considering its social responsibility are listed below.

(1) Should it operate subsidiaries in other countries where there is overt racial discrimination?

(2) Should it purchase abroad when the comparable British product is slightly more expensive?

(3) Should it provide safe systems of working beyond those required by the Health and Safety at Work Act, 1974?

(4) Should it open a new factory in an area well-known for the militant attitude of its labour force but with persistently high unemployment?

(5) Should it fully protect the purchasing power of its ex-employees' pensions from the effects of inflation?

(6) Should it cease production of one of its products which, although not banned by law, can seriously harm the health of those who use it?

2.4.4 Survival

This objective is strongly present in most companies but it may not be widely acknowledged, particularly when profit and growth targets are being achieved. However, it is usual for unwelcome take-over bids to be resisted vigorously by the board of directors.

A sound policy of product development (see section 5.11) is essential for longer-term survival. If a company anticipates that its traditional markets will contract, it should seek new ones. If it anticipates that its product range will not be compatible with new

demand trends, it should respond quickly by introducing new designs.
If it fears that survival depends too much on a narrow range of products, it should diversify.

Survival also depends on having adequate succession for positions
of responsibility. Therefore the selection and training of staff
for promotion should be an important part of company planning.

Another factor influencing the long-term survival of a company is
keeping its manufacturing methods up-to-date. This can be assisted
by ensuring that there are regular and ample funds available to
purchase new equipment.

In times of economic recession, survival becomes a prime objective
in many companies. Faced with lower demand, production capacity has
to be cut and overheads pruned. It is also likely that profit will
have to be reduced and work may have to be taken at a price that
barely covers marginal costs (see section 8.6). These measures will
help to maintain at least a nucleus of the work-force and could
assist the company to survive until conditions improve.

2.4.5 Interaction of Company Objectives

It has already been indicated that companies do not pursue a single
objective to the exclusion of the rest. Emphasis and priorities in
objectives will change from time to time with variations in trading
conditions and changing company aspirations. Measures that are
necessary to promote one objective can affect others. For instance,
when a major new product-development programme is undertaken it will,
if successful, assist growth and survival but could reduce profit in
the short term due to the retooling and launching costs.

Corporate planning attempts to determine the best mix of company
objectives and to put them into a master plan for the organisation.

2.4.6 The Corporate Plan

Companies can either react to external changes as they occur or
follow a plan that attempts to influence events to their advantage.
If they wish to adopt the second course they must decide where they
want to go and then make plans to get there.

It would be unsatisfactory if the marketing, personnel, finance,
production, and engineering functions were to make their separate
plans; co-ordination is necessary and this can be achieved by means
of the corporate plan. A single formal plan not only ensures that
departmental plans are co-ordinated but gives managers a broader view
of the activities of the company as a whole and provides them with
a framework within which they can contribute to the firm's strength
and prosperity. The usual period covered by the corporate plan is
one year, although medium-term plans covering a five-year period and
long-term plans looking forward for, say, ten years, are sometimes
prepared.

Four stages can be identified in the production of a corporate
plan; they are, the situation audit, setting objectives, preparing
the plan, and implementing the plan.

Situation Audit

This preliminary stage is necessary to decide where the company
stands. Periodic auditing at, say, annual intervals reveals company
strengths and weaknesses and detects changes in the external environment relevant to the company's future.

The internal operation of the company should be reviewed; in particular failures in performance should be detected and examined. The market into which the company is selling should be looked at to reveal changes in market size, market share and competitive activity. At the same time trends in customer preferences and buying habits should be examined. A study should be made of the general social and economic environment; this should include actual and probable changes in taxation, tariffs, and legislation, together with trends in public opinion likely to affect the firm.

Setting Objectives
Here the company sets its course. Corporate objectives have already been discussed in this section and an appropriate selection is made with relevant priorities. Objectives should be stated precisely and defined closely to ensure that the frame of reference remains unchanged. An example of a major objective could be increasing profitability from 18 to 20 per cent and an example of a minor objective could be to reduce the operator waiting-time on the press section from six to four per cent of the clocked hours - both over a period of one year.

Compiling the Plan
This stage considers how to achieve the objectives. A strategy has to be decided upon and agreed at the highest level. Frequently this is done only after a number of alternative strategies have been considered and discarded. After the general strategy has been agreed, a master plan is prepared. This covers the total operation of the company over the planning period. There are a number of facets to this plan. The financial one is concerned with the projected balance sheet and profit and loss account and with capital requirements. Resources, both human and material, are planned, together with a product-development programme. The plan should be broken down into departmental and sectional targets. Many of these can be dealt with by incorporation in the annual financial budget which will be discussed in chapter 8. Other aspects involving human resources can become part of manpower planning, to be discussed in chapter 7.

Implementing the Plan
Individual responsibilities must be allocated for the achievement of the network of targets generated by the corporate plan. Regular reviews of achievement against target should be made and variances from the plan established. These variances can be the subject of monthly review meetings. Changing circumstances will necessitate periodic updating of the corporate plan; a frequent cause of updating is when actual sales differ from those that have been forecast.

2.5 LOCATION OF INDUSTRY

2.5.1 Historic Development
In the nineteenth century, British industry was centred around its coalfields. Coal was required for iron and steel production and for the steam-engines that powered the factories and mills. Industry attracted labour and large conurbations rapidly developed in the Midlands, in the North of England, in South Wales, and in the Clyde Valley. A skilled labour force was developed, specialised commercial services were built up, and good railway communication and port facilities were constructed.

In the twentieth century the electric motor has replaced the steam-
engine as the method of driving machines in factories. Electricity
is universally available throughout the country and proximity to
markets has replaced nearness to coalfields as a major consideration
in industrial location. London as the main centre of population has
become a magnet for industrial development and the 1920s and 1930s
saw the beginning of an industrial drift to the South. The Midlands,
helped by their skilled and adaptable labour force and their central
location also experienced an industrial expansion and became the
centre of the motor industry. If there had not been government
intervention, starting with the Special Areas Act of 1934, the south-
ward drift of industry would have accelerated and the economic
balance of the country would have been destroyed.

2.5.2 Factors Affecting Industrial Location
Let us consider some of the factors influencing the location of
industry.

Proximity to Markets
Although delivery costs will reduce with closeness to markets, much
will depend on the type of product being sold. With relatively low-
cost bulky products such as cement, sand, and gravel, market-proxim-
ity is important. As the value/bulk ratio increases, the location of
the market becomes less significant. For instance, a manufacturer of
jet engines need not be located near to its customers. Nearness to
markets is also advantageous to companies offering a sub-contract or
component-manufacturing service as this promotes better liaison and
a more rapid response to customer requirements.

Proximity to Sources of Production Materials
The importance of this factor will depend also on the type of product. The lower the value added by the manufacturing process and the
more bulky the materials of production, the more important it is that
the plant should be near to the materials it uses. For instance,
blast furnaces should be near to deposits of coal or iron ore and
cement works should be near to deposits of chalk or limestone.

Availability of Labour
An adequate supply of suitable labour can be an important consideration. This factor obviously assumes greater importance in times of
near-full employment. Except in sparsely populated areas, adequate
supplies of unskilled labour are not normally difficult to obtain.
Recruitment of skilled labour of the appropriate type can, however,
create greater difficulties. Most large industrial conurbations,
however, offer a wide range of ready-made skills and good facilities
for the training and retraining of employees.

Labour Attitudes
Some areas have acquired a reputation for their militancy and for
their reluctance to accept change. It is hardly surprising that
many companies are unwilling to move to these areas.

Communications
The availability of good road, rail, sea, and air communications are
factors influencing industrial locations. Continental Europe is
Britain's major export market and the South East of England is more

favourably situated than other regions with respect to communications with the Continent of Europe.

Environmental Factors

Unskilled labour tends to be immobile but this is not so with technical and management personnel. The shopping, recreational, and educational facilities, coupled with the climate and landscape of the region can strongly influence the quality of applicants when vacancies for senior staff are advertised. Cost of housing is another factor although some companies will help staff financially if they move to an area of high housing costs.

Government Policy

The government has been attempting to influence the location of industry for almost 50 years by attracting new industrial development to designated areas of higher-than-average unemployment. In the past they have used the policy of the 'carrot and the stick'. The carrot was in the form of considerable financial inducements to set up factories. The stick was the power to refuse permission for any significant new industrial development except in designated areas. Today, although the financial inducements remain, it is no longer necessary to obtain an industrial development certificate from the Department of Trade and Industry before new building can start.

FURTHER READING

Bale, J., *Location of Manufacturing Industry* (Oliver and Boyd, Edinburgh, 1981).

Betts, R.J., *Business Economics for Engineers* (McGraw-Hill (UK), Maidenhead, 1980).

Burton, J., *The Job Support Machine: A Critique of the Subsidy Morass* (Centre for Policy Studies, London, 1979).

Glew, M.T., Watts, M.G., and Wells, R.M., *Business Organisation and its Environment*, Books 1 and 2 (Heinemann, London, 1979).

Norbury, P. and Bownas, G., *Business in Japan* (Macmillan, London, 1980).

Reekie, W.D., *Macroeconomics for Managers* (Phillip Allen, Oxford, 1980).

3   Management Organisation

A one-man business depends entirely on the efforts of its owner but
as soon as others are employed their work must be organised. This
chapter is concerned with the type of formal organisation in which
engineers are likely to work; it also indicates how the work of
individuals can be integrated to achieve specific objectives.

3.1 FORMAL ORGANISATIONS

Discussion and speculation on the organisation of people can be found
in the writings of the organisational theorists. There are many
schools of management thought; eight have been identified by H.
Mintzberg in *Nature of Managerial Work*.

There has been a reluctance of one school to listen to another and
later schools often fail to take sufficient account of the contribution made by earlier schools. It is, therefore, not surprising
that most companies evolve their management structure with little
direct reference to the jungle of opinion existing in organisation
theory.

There is no formal list of schools of management thought; four of
the better-known ones will be briefly described in this chapter.
They are the classical school, the behavioural school, the quantitative school, and the systems school. Additional information on
the classical and the behavioural schools can be found in section
3.2, where the work of the pioneers of management thought will be
reviewed.

3.1.1 The Classical School

The work of this school is chiefly concerned with the structure and
operation of formal organisations. A division of labour is specified
by management who determine the rôle of each job and allocate work.
There is emphasis on the production function, where standard tasks
are carefully studied and attempts made to find the single best
method of working.

Subsequently, work measurement is used to monitor performance and
for incentive schemes. If there are tasks that cannot be standardised, uniform procedures are adopted. There is little attention to
the feelings of the employees doing the work; they are merely fitted
into carefully designed work situations. Fayol, Taylor, and Weber
were major contributors to this school.

3.1.2 The Behavioural School

This approach to organisation is people-centred and was a reaction
against the highly mechanistic classical school. The behavioural
school is also known as the human relations school. Most of the
contributors have come from the ranks of psychologists and sociol-
ogists. Their work has been concerned with learning, motivation,

group behaviour, and leadership in an industrial environment.

The behavioural approach to management does not have organisational rules, as does the classical school, but importantly it tries to ensure that employees are treated with consideration rather than as extensions of the machines that they operate. It emphasises the need to use employee-oriented democratic forms of management that are supportive of the individual employee.

Many consider that this school began with Mayo in the 1930s, although Follett was advocating a more democratic form of management well before that date. Other leading contributors include McGregor, Argyris and Likert.

3.1.3 The Quantitative School

This school stemmed from the operations research work done by inter-disciplinary teams to solve problems posed by land, sea, and air operations during the Second World War. Here mathematical processes and models were used to assist in problems involving the acquisition, disposition, and utilisation of resources.

A similar approach proved to be of value when applied to industrial problems; among the techniques used and developed were linear programming, network analysis, decision trees, and simulation.

The availability of a number of well-developed decision models is of considerable value to industrial managers and helps them to mini-mise guesswork in decision-making. The validity of the answers provided will depend on an appropriate choice of mathematical model and the accuracy of the input data employed.

3.1.4 The Systems School

This relatively recent approach considers an industrial organisation as a system. The system is made up of a hierarchy of subsystems; the operation of the company can be considered as the system, with production, marketing, financial, and other subsystems acting as 'building blocks'. The subsystems are interrelated and should contribute to the purpose or objective of the system.

Most production systems consist of loosely coupled subsystems. Complete integration may neither be desirable nor possible, although where the possibility exists, as in the case of a modern steelworks, considerable cost savings may be achieved by integration. A common symptom of the lack of integration is buffer stocks; these decouple one subsystem from its neighbour and allow temporary difficulties in one subsystem to be absorbed without affecting other subsystems.

All systems operate in an environment which, for a manufacturing company, will include customers, suppliers, trade unions, competitors, and the government. Although systems are affected by the environment in which they operate, they have limited ability to modify their environment.

Time-lags are inherent in all systems and can result in poor response to changes in input. When the output of the system is monitored, as a control on inputs to the system, the time-lags can reduce the effectiveness of controlling-action and in some instances produce instability.

Brief consideration will now be given to the inputs, transformation, output, and control of a manufacturing system.

Inputs
These can be subdivided into basic and complementary inputs. In a

production system, the basic inputs are those that appear in the finished product such as material and bought-out parts.

Complementary inputs are those that are used to transform the basic inputs into the finished product; they comprise resources such as people and machines as well as information and instructions needed to operate the system.

Transformation

Manufacturing is not normally a single-stage operation, as indicated in some simplified system diagrams, but is usually a complex multi-stage procedure. It may be possible to show the interrelationship of manufacturing operations graphically in network diagrams; the construction of network diagrams is described in section 10.4.

Outputs

These, like inputs, can be divided into basic and complementary components. The basic outputs are the goods or services produced; complementary outputs include product quality and cost data, plus social outputs such as work satisfaction or frustration.

Control

The purpose of control is to ensure that the system transforms input into output in accordance with the objectives of the system. This is achieved by establishing control-points and feeding back information designed to keep the system in control. Feedback control uses inspection and other operating information from within the system and possibly from the system's environment to keep the system in control.

A much simplified diagram of the operation of a manufacturing system is shown in figure 3.1.

3.1.5 A Unified Theory of Management

It will have been seen that the various management schools have little common ground. They tend to be a collection of ideas and theories generated by people having different interests and backgrounds. Because of this divergence it seems unlikely that a unified theory of management will emerge by integrating present theories. It is also unlikely that one of the existing schools will be accepted as the correct one.

3.1.6 The Contingency Theory

This theory proposed that the most appropriate organisation structure, either classical or behavioural, will depend on the type of company. Indeed, within a single company, different organisational approaches are appropriate; for instance, the classical approach in the factory with the behavioural approach in the research and development sections.

The Woodward Studies

These were organised in the 1960s by Joan Woodward of Imperial College, London. She studied a large number of British companies which were categorised according to the type of production used: (a) jobbing and small-batch production, (b) large-batch and flow production and (c) continuous-process production. It was found that in large-batch and flow production there was a clear chain of command, highly specialised jobs, centralised control, and supervisors responsible for a large number of operators. However, in jobbing, small-

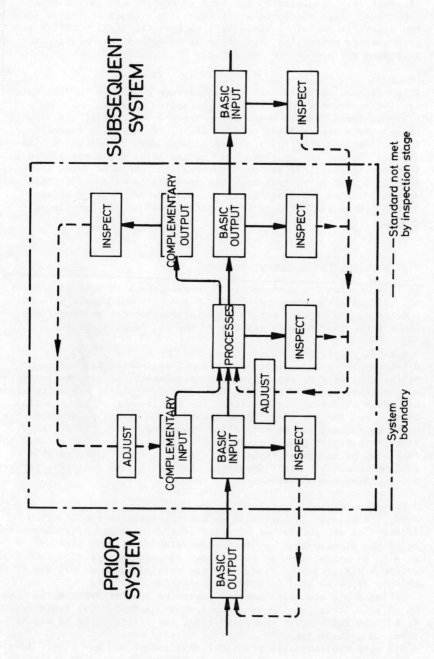

Figure 3.1 Elements of an industrial system

batch, and process production, the chain of command was less clear, the work less specialised, and each supervisor controlled fewer employees. In consequence, large-quantity production tended towards the classical school of management whereas, in small-quantity and process-type production, the style of management was similar to that advocated by the behavioural school.

The Burns and Stalker Studies

These studies of 20 British companies were conducted to find which type of management approach was best suited to each.

It was found that in companies where the environment was slow to change there was emphasis on efficiency, the work was specialised, and well-established rules and procedures were in operation. In this type of company the approach was 'mechanistic' and corresponded to that of the classical school. However, in companies where there was an environment of rapidly changing products or processes, there were few rigid rules and flexibility was encouraged. The management approach here was 'organic' and corresponded to the behavioural school.

3.2 PIONEERS IN MANAGEMENT THOUGHT

In this section the contributions to management thought of some of the leading writers are reviewed. Space will not permit more than a brief review of each but further information can be found from the books listed for further reading at the end of this chapter.

3.2.1 H. Fayol (1841-1925)

Henri Fayol was an engineer and manager who worked for a French mining company. He considered that management was an administrative science that could be taught. Fayol is particularly remembered for his ideas on formal organisation, the development of organisation charts, and the use of job descriptions. He strongly advocated the unity of command and direction and recommended that the authority given to managers should match their responsibility. To relieve the work load on managers, Fayol suggested that they should be able to call on specialist staff for help; this specialist staff was to have no operating responsibilities.

3.2.2 F.W. Taylor (1856-1915)

Frederick Winslow Taylor was a mechanical engineer who is often referred to as 'the father of scientific management'. He is reputed to be the first person to use a stop-watch to measure work and to couple the performance of these measured tasks with a wage incentive scheme. His system of time study was relatively crude but was greatly improved by C.E. Bedaux who introduced rating.

Taylor wrote several important papers on factory organisation and on metal-cutting for the American Society of Mechanical Engineers. In 1895 he was largely responsible for the introduction of high-speed steel, a greatly improved alloy for cutting metal.

His work had immediate practical application and was widely adopted in the United States. His ideas on work measurement and incentives greatly increased productivity but were often badly applied and caused considerable controversy and friction.

3.2.3 M. Weber (1864-1920)

Max Weber was a German sociologist who worked in hospital administration during the First World War. His theory of bureaucracy had
considerable influence on thinking about formal organisation.
Weber's bureaucracy was an ideal organisation, setting standards
against which other organisations could be measured. It was a model
of efficiency and quite different from the red tape and sloth that
many now associate with the word bureaucracy.

  Weber proposed a well-defined and powerful hierarchy with clear
lines of responsibility and a high degree of job specialisation. He
advocated a system of rules covering the rights and duties of employees. Managers were to be recruited on the basis of ability and
technical knowledge; their relationships with subordinates were to
be on a formal and impersonal basis.

  Weber's theory of organisation proved to be too rigid and unbending
and was in contrast to that of Fayol which was a more flexible and
practical approach.

3.2.4 M.P. Follett (1868-1933)

Mary Parker Follett was an American political scientist who examined
the best in management practice both in the United States and in
Britain. She advocated that responsibilities should be shared and
that authority should be diffused throughout the organisation. When
standard practice was followed, she recommended that it should be
determined by joint consultation between management and those involved and not arbitrarily imposed by management. Follett also proposed that co-ordination between departments should be achieved by
the joint planning of the departments concerned.

3.2.5 F.B. Gilbreth (1868-1924)

Frank Bunker Gilbreth, an American and a contemporary of Taylor,
introduced and developed motion study. Motion study is a close
analysis of the movements involved in doing work with the object of
making them more efficient. Gilbreth was assisted in his work by
his wife Dr Lillian Gilbreth, a psychologist. The work of Taylor
and the Gilbreths has proved to be complementary and now forms the
basis of the subject of work study.

3.2.6 G.E. Mayo (1880-1949)

Elton Mayo and his associates Roethlisberger and Dickson focused
attention on the effect of group attitudes on employee productivity.
Mayo was an Australian and a professor at Harvard University.
Roethlisberger was also from Harvard while Dickson worked for the
Western Electric Company. They based their conclusions on studies
carried out in the 1920s and 1930s at the Hawthorne Works of Western
Electric in Chicago.

  Initial studies were concerned with the effect of the physical
environment on output. One of the findings was that the output of
coil winders and relay assemblers increased as the level of light
intensity supplied to the work-place increased. However, output did
not reduce, as expected, when light intensity was subsequently
reduced. The investigators concluded that it was the good personal
relationships established with the group of employees that had spurred
them to greater efforts with each change made by the investigators.
An improvement in operator performance that results from direct observation and management interest is often referred to as the
Hawthorne effect.

In another set of studies it was found that a group of experienced male operators, involved in wiring banks of relays, formed between themselves an intricate social structure that set output norms. Despite a financial incentive, any group member who exceeded the higher norm or failed to reach the lower norm was forced into line by the unofficial group leaders. This study indicated the importance of the relationships between group members in determining levels of output.

The Hawthorne studies did not show that financial rewards can be disregarded when considering employee motivation. Inadequate pay creates strong dissatisfaction and a monetary reward that increases with effort can act as a positive motivating factor for many employees.

3.2.7 R. Likert (1903-   )

Renis Likert, an American social psychologist, made a study of the supervisor and his influence on productivity. He distinguished between job-centred and employee-centred supervisors. The job-centred supervisor looks closely at the work to be done and considers the people who do the work as instruments for achieving work targets. He delegates as few decisions as possible and closely supervises the work as it is done. The employee-centred supervisor displays greater personal interest in his staff, using participative techniques to achieve a high level of group loyalty. He sets high goals for his group of employees and lets them get on with their tasks; he is, however, always available to give support when needed.

In general, higher productivity is achieved by an employee-centred supervisor. The amount of supervision that is desirable will vary with the task being performed; on non-routine work most employees welcome close guidance.

3.2.8 D. McGregor (1906-1964)

Douglas McGregor also was a social psychologist who became Professor of Management at the Massachusetts Institute of Technology. He is particularly remembered for his Theory X and Theory Y of human behaviour.

Theory X assumes that people dislike work, are lazy, lack ambition, resist change, and are indifferent to the aspirations of the company that employs them. Close external control will be needed if employers are to obtain a fair day's work from their employees. Theory Y assumes that people want to work, can be taught to accept responsibility, and will exercise self-control and initiative.

McGregor did not accept that people will naturally behave according to Theory X and considered Theory Y to be the more realistic. Employees should be helped to develop their inherent potential by sympathetic management who encourage employee participation in decision-making and show appreciation of employee efforts. If this is done, employees will identify closely with the objectives of the firm and greater productivity will result.

3.2.9 W.D. Brown (1908-   )

Wilfred Brown was a practising manager and one of the pioneers of professional British management. He joined the Glacier Metal Company in 1931 and was its Chairman and Managing Director from 1939 until 1965. For a considerable part of this time he collaborated with the social scientist, Elliot Jacques, in the Glacier Project; this project was an exhaustive study into the organisation and management of an industrial company.

Brown tried to manage his factory as a cohesive society. He established advanced forms of works councils that helped to encourage participation at all levels. In his search for the best form of executive system he proposed the idea of 'operational work' which was the essential profit-earning activities of the company; Brown identified operational work as developing, manufacturing, and selling the company's products. Other work was seen to be in a supporting rôle to operational work and operational managers at each level were provided with their own supporting staff of specialists.

Wilfred Brown wrote a number of books that explained his ideas on industrial management. He received a life peerage in 1964 and is now Baron Brown of Machrihanish.

3.2.10 A.H. Maslow (1908-1970)

Abraham Maslow proposed that people had a hierarchy of needs. When they had satisfied the lowest level they proceeded step by step trying to satisfy higher needs until they reached a personal ceiling beyond which they did not rise.

Maslow's hierarchy of needs are listed below with the first level being the lowest.

1st	Food and shelter	(Keeping alive)
2nd	Security	(A stable and ordered life)
3rd	Membership	(Acceptance by others)
4th	Esteem	(Prestige, success)
5th	Self-fulfilment	(Realisation of full potential)

Work in industry should, without great difficulty, satisfy the first and second levels. Many companies attempt to provide work systems that satisfy the third and fourth levels but little can be done about the fifth level except for relatively few employees.

3.2.11 P.F. Drucker (1909-)

Peter Drucker moved to London in the 1930s to escape Nazi persecution; afterwards he went to New York where he worked as a business consultant. His writings on management are of direct use and have been readily accepted by practising managers.

One of his major interests has been the organisation of large corporations. He suggested that activity analysis, decision analysis, and relations analysis should be used to determine the type of organisation that would suit the objectives of a particular company. Activity analysis considers the work that needs to be done and how it should be grouped so that the objectives of the firm can be achieved. This is in contrast to the ready acceptance of the work-grouping of conventional organisation charts. Decision analysis indicates where decisions should be taken. Company decisions can be analysed by considering for how long they are likely to affect the company, whether or not they depend on ethical values, whether they are of a routine or non-routine nature, and the effect of the decisions on other parts of the company. The more far-reaching, long-term, non-routine, and politically sensitive the decision, the higher up the organisation it should be taken. Relations analysis improves the understanding of the relationship that each post of responsibility has with others in the organisation. It considers the

contribution made by each post and its upward, sideways and downward
responsibilities. Apart from improving the organisation structure
of the company, organisation analysis can be helpful in the selection
of managers since it gives a more complete picture of the post to be
filled.

Drucker suggests three structural requirements for a company. The
first is that it should be organised to match the objectives of the
firm. The second is that it should have as few layers of management
as possible. The third structural requirement is that it should
enable future senior managers to be trained while they are still
relatively junior and while the consequences of any incorrect decision are not too disastrous. To achieve these structural requirements it is desirable that larger companies should be decentralised
on a product basis with each product division being responsible for
its own operation and profitability. If this type of organisation is
impossible then the company should be decentralised around a function
with each function representing a distinct phase of the business.

3.2.12   C. Argyris (1923-   )

Chris Argyris, a social scientist and Professor of Industrial Administration at Yale, was concerned with the causes and effects of the
frustration and aggression that occur in industrial organisations.

Argyris proposed that if employees cannot find a meaning in their
work they will become frustrated and be placed in conflict with
management. At higher levels this can be seen as mistrust and unwillingness to accept risks while at lower levels it is shown as
apathy, dependence, and collective distrust. Argyris proposed that
job enlargement should be used to help employees to use their potential more fully. He advocated an employee-centred democratic management which allowed employees to participate in decisions that
affected them.

3.2.13   F. Herzberg (1923-   )

Frederick Herzberg *et al*. analysed factors that motivate employees
at work. These factors are divided into hygiene factors and motivating factors. Hygiene factors are unlikely to create employee
satisfaction yet cause dissatisfaction if they are absent. They include working conditions, quality of supervision, group relations,
and monetary reward. Motivating factors provide satisfaction and
include recognition of responsibility, and achievement.

For motivating factors to be effective, the hygiene factors should
also be present. Hygiene factors satisfy the lower levels of
Maslow's hierarchy and motivating factors satisfy the higher levels.

3.3   ORGANISATION CHARTS

These charts show the formal organisation structure of a company and
indicate who is responsible to whom. Conventional organisation
charts consist of boxes joined by lines; the boxes have the job
titles and possibly the post-holder's name written in them while the
lines show lines of nominal responsibility. The more important the
job, the nearer it should be to the top of the chart but, when drawing organisation charts, it is prudent to add a footnote that tasks
of equal seniority are not necessarily shown at the same level.

An example of an organisation chart which shows a chief executive,
his assistant, and five departmental managers appears in figure 3.2.
Although the assistant to the chief executive is shown in this chart

above the departmental managers he is not necessarily senior to them.
Organisation charts are not static but change as the company changes
in size or alters the direction of its business.

It is unusual for organisation charts to be drawn showing every
person employed; they are normally restricted to management levels.
In large organisations, separate charts are often drawn for the
management structure of each major subdivision.

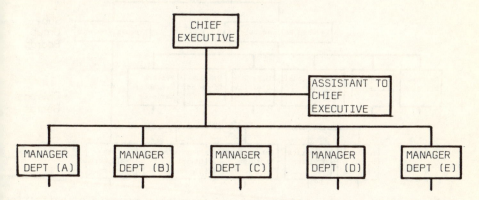

Figure 3.2 An organisation chart showing senior management

3.3.1 Primary Grouping of Activities
A fundamental consideration in company organisation is the primary
subdivision of its activities. There is likely to be a degree of
compromise when choosing the most appropriate primary subdivision.
Some of the possible choices are described in this section.

Geographical Region
Where a company has a widespread international operation, this is
likely to be reflected in its organisation chart. A number of regional chief executives may be appointed with each responsible for
operations in his own region and having considerable freedom to make
important decisions. This type of organisation enables companies to
adapt readily to local conditions and to make decisions more rapidly.
Major oil and other multi-national companies often organise themselves in this way.

Product Grouping
Large multi-product companies frequently find an organisation based
on a grouping of similar products to be the most effective. Perhaps
the best-known British example of this type of organisation is
Imperial Chemical Industries. This company has a number of self-
contained divisions each producing a group of similar products, for
instance, paints.

Some companies have adopted a mixture of product and regional subdivisions. The Beecham Group is divided into two main sub-groups:
pharmaceutical and non-pharmaceutical products. Below this, the
divisional boards are largely a mixture of regional and product subdivisions. The organisation structure of the Beecham Group is shown
in figure 3.3; the boxes in this figure represent a board of direc-

tors and not individual managers. There is interlocking membership
between board levels.

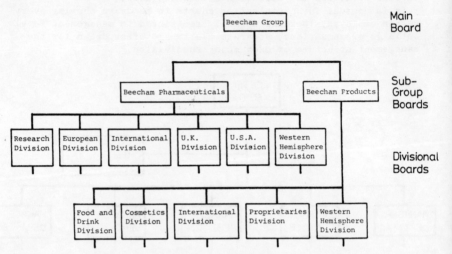

Figure.3.3 A mixture of product and geographic grouping of activities

Functional Grouping

This popular form of organisation is illustrated in figure 3.4 and
is the primary method of division of responsibility of most larger
engineering companies operating on a single site. However, when used
as the primary grouping for a multi-site operation it inhibits decentralisation and makes it difficult to isolate the performance of individual sites.

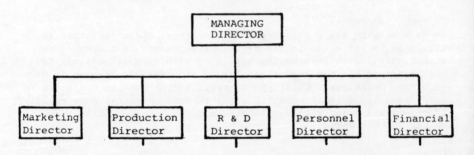

Figure 3.4 Primary subdivision by function

Project Grouping

An organisation structure of this type is suitable for companies that
have a strong project orientation such as the aircraft or construction industries. A project grouping consists of a number of self-
contained functional groupings each under a project manager who
reports to the chief executive of the company. Project grouping
clarifies responsibilities and helps employees to build up a loyalty
and commitment to the project on which they work.

A more complex project-based structure can exist as a matrix. This structure cuts across the dictum 'one man one boss' since functional managers are responsible to both their project director and their functional director. A typical matrix structure for a project-orientated company is shown in figure 3.5. Matrix structures can create conflict and divided loyalties but work well if goodwill exists. They combine the advantages of project-orientated management and central support for functional managers.

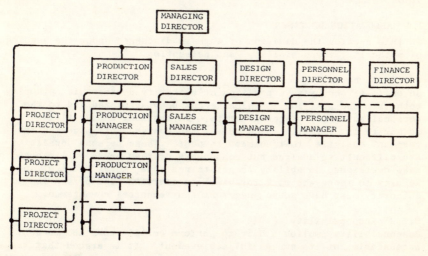

Figure 3.5   Matrix structure with primary subdivision by project and by function

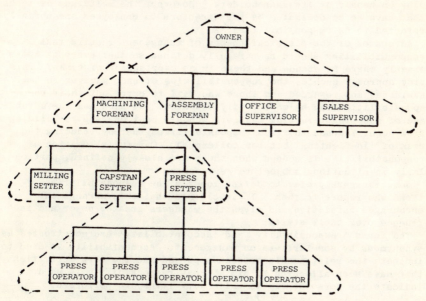

Figure 3.6   Primary subdivision by process also showing interlocking management groups

Process Grouping

Here the primary division is by the clerical or production processes.
It is suitable only for small companies although organisation by
process is widely used as secondary or tertiary subdivisions in larger
companies. A typical division by process in a small engineering
company is shown in figure 3.6. The dotted lines on this organisation
chart indicate three typical interlinked management groups; the inter-
linking of management groups is typical of any management organisation.

3.4 ALLOCATION OF TASKS

In this section the division of work is discussed under the headings
of authority, responsibility, and delegation.

3.4.1 Authority

An engineer occupying a position of authority derives his authority
both from the position to which he has been appointed and to the
personal authority he can generate. In the long run, official
authority is of little value unless the post-holder possesses the
personal qualities that command respect. These personal qualities
are difficult to analyse but components of them are a willingness to
make decisions, an ability to get things done, moral courage, and
loyalty to those who work for him. In addition, the person in
authority must have sound judgement and technical competence.

3.4.2 Responsibility

Responsibility implies a duty to perform certain work and to be
accountable for its successful achievement. It is argued that tasks,
not responsibilities, can be delegated and, of course, the board of
directors shoulder the ultimate responsibility for running the company on behalf of its shareholders. However, the multitude of tasks
that have to be delegated by the directors to employees are normally
referred to as responsibilities.

Advocates of the classical theory of management require that
responsibilities should be carefully defined so that everyone knows
exactly where he stands and there is no overlapping of tasks. This
tidy approach enables the responsibilities of each manager to be
evaluated and rewarded and those who fail to carry out their responsibilities can be more easily detected. It also has the advantage of checking thrusting managers who are bent on 'empire building'
from acquiring unauthorised responsibilities thereby reducing the
risk of 'in-fighting' between colleagues. Careful allocation of
responsibilities is needed when these are closely defined, particularly in situations subject to rapid change.

Some managers prefer to define loosely the responsibilities of
those who report to them. It is claimed that this arrangement
encourages initiative and gives the managers greater power and influence over their staffs.

The terms 'responsibility' and 'accountability' are not treated as
synonymous by some writers on management. Accountability is used to
indicate the relationship between a person and the work activities
that have been allocated to him whereas responsibility is used to
indicate the personal way in which activities are performed.

3.4.3 Delegation

Delegation is the allocation of tasks to employees. For delegation

to be effective, the person to whom the tasks have been delegated
should have a clear idea of what he is meant to achieve, should be
willing and able to do the work, and should possess the necessary
authority.

A balance should be maintained between over- or under-delegation.
With too great a degree of delegation, senior management can lose
control and unsatisfactory decisions may be made by inexperienced
junior staff. Under-delegation results in overworked senior management with inadequate time to consider important decisions.

Although Fayol's proposal that the responsibility delegated to a
person should correspond with his authority is still valid, it should
not be too rigidly applied, otherwise managers might adopt an outlook
that is too restrictive and neglect broader issues of company
interest.

3.5 SPAN OF CONTROL

Span of control refers to the number of people who report directly
to a particular person. In a company, the larger the span of control
the fewer the layers of management. It is impossible to state the
optimum number of employees who should report to a manager. The
effective span of control is much larger when well-organised standard tasks are being performed. Complex tasks of great variability
may have an effective span of control of two or three persons only.

The fewer the number of management layers the better the communication between the top and bottom of the company. Communication
will be discussed in chapter 4. The effect of the span of control
on the number of layers of management in two companies having an
equal number of operators is shown in figure 3.7.

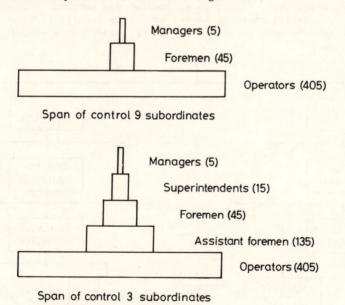

Span of control 9 subordinates

Span of control 3 subordinates

Figure 3.7 Effect of the span of control on the number of layers of
            management

3.6 CO-ORDINATION

The organisation structure is simply a framework illustrating jobs and their relationships. A detailed job description is helpful in indicating the extent of the duties and responsibilities of a particular job. Organisation charts and job descriptions do not ensure that people do their jobs properly and that difficulties and disputes do not arise.

Co-ordination is needed to help ensure that duties are effectively performed. Some co-ordination is achieved voluntarily with employees seeing the need and using their initiative to get on with work in the best interests of the company. The level of voluntary co-ordination is high in small organisations as here people can see their rôle more clearly.

Co-ordination is usually achieved by laying down guidelines and providing decision rules that indicate what should be done in particular circumstances. Co-ordination can be provided in the factory by supervisors giving instructions to operators and by the use of documents such as operation layouts and material requisitions. The use of computers for data processing has done much to promote company-wide co-ordination because they rapidly analyse and process data.

Co-ordination within a department is usually easier to achieve than co-ordination between departments. However, in a manufacturing company, inter-departmental co-ordination is vital if the company is to operate successfully. This inter-departmental contact is referred to as a staff relationship as distinct from the hierarchical line relationship that exists within a department. The staff relationship that is shown in figure 3.8 could occur when a machine-setter is advised by an inspector to reset a machine that has been producing sub-standard work. Staff relationships are advisory and not order-giving. Unwillingness to accept advice is usually referred for settlement to more senior levels in the departments concerned.

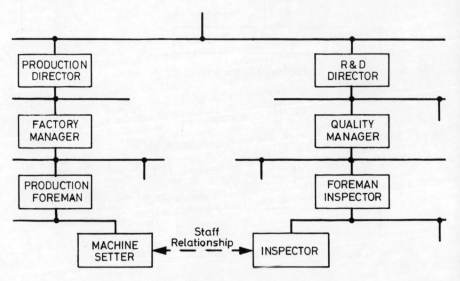

Figure 3.8 Part of an organisation chart showing a staff relationship

FURTHER READING

Argyris, C., *Increasing Leadership Effectiveness* (Wiley, New York, 1976).

Brown, W. and Jacques, E., *Glacier Project Papers* (Heinemann, London, 1965).

Drucker, P.F., *Practice of Management* (Pan Books, London, 1968).

Fayol, H., *General and Industrial Management* (Pitman, London, 1949).

Gilbreth, F.B. and Carey, E.G., *Cheaper by the Dozen* (Heinemann, London, 1949).

Herzberg, F. *et al.*, *Motivation to Work* (Wiley, New York, 1959).

Likert, R., *Human Organisation: Its Management and Value* (McGraw-Hill, New York, 1967).

McGregor, D., *Human Side of the Enterprise* (McGraw-Hill, New York, 1960).

Maslow, A.H. (Ed.), *Motivation and Personality* (Harper Row, New York, 1970).

Mayo, G.E., *Social Problems of Industrial Civilisation* (Routledge and Kegan Paul, London, 1975).

Mintzberg, H., *Nature of Managerial Work* (Prentice-Hall, Englewood Cliffs, NJ, 1980).

Taylor, F.W., *Principles of Scientific Management* (Greenwood Press, Westport, Conn., 1947).

Woodward, J., *Industrial Organisation: Theory and Practice* (Oxford University Press, Oxford, 1980).

4 Communication

4.1 THE NEED FOR COMMUNICATION IN INDUSTRY

An engineer should be a good communicator. In his own organisation
he must communicate upwards to those he is responsible to, laterally
with colleagues, and downwards to those who work for him. He is also
likely to communicate externally with customers and suppliers and to
write papers for conferences and journals.

Communication within industry has assumed greater importance because employees expect to be kept better informed and to participate
in decisions that directly affect them. Good communications can
create a spirit of unity and co-operation within a company, as well
as help to avoid misunderstandings and industrial unrest.

In small companies, lines of communication are short and people
can see their rôle clearly. As the size of the organisation grows,
more attention must be paid to formal methods of communication.

4.2 METHODS OF COMMUNICATION

4.2.1 To Individuals Face to Face

Supervisors and managers frequently use this method of communication
to pass on orders and instructions.

Face-to-face communication helps to establish the manager or supervisor as the leader of his group. It is flexible as the words used
can be adapted to the particular group and to its mood. It also
enables questions to be asked both ways, thereby helping to ensure
that the message has been understood.

4.2.2 Through Representatives

Here management talks to the elected representatives of employee
groups. Time is saved if this method of communication is used but
the messages may become distorted in transmission by the representatives and the position of the supervisor as the leader of the work-
ing-group may be undermined.

4.2.3 By General Announcement

This method enables a large number of people to be informed rapidly.
The announcement may be made over a loudspeaker system, at a mass
meeting of employees, or in a written notice posted on notice-boards.
There is no opportunity for questions and hence there is less likeli-
hood of the message being understood when an announcement is made by
loudspeaker or posted on notice-boards. With mass meetings, the
possibility of questions exists but the number asked is usually small.

4.2.4 Via the 'Grapevine'

In this unofficial method of communication, information is rapidly
passed by word of mouth. The grapevine flourishes in large companies

where communication has been neglected. Sources of information for
the grapevine vary from listening to other people's conversations to
staff improperly divulging confidential information. Grapevine
information is usually true, or partly true, but often it is unchari-
tably slanted. It can cause attitudes to harden against management
policy prematurely and on occasions can result in unnecessary dis-
tress.

4.3 DOWNWARD COMMUNICATION

4.3.1 Major Changes
It is important that information concerning major changes should be
carefully handled and fully understood. For instance, if a board
decision has to be passed right down a large organisation, the
director responsible for each function will call a meeting of his
departmental managers. Each departmental manager will then hold
meetings with his supervisors who in turn will hold meetings with
the group of employees for whom they are responsible. The nature of
the meetings will be modified to suit the audience and questions
should be encouraged. A written statement of the change should be
available to the chairman of every meeting to avoid distortion of
the message as it passes from stage to stage.
  The release of sensitive information must be carefully planned
with precise information given concerning when and to whom it should
be released.

4.3.2 Regular Briefing
Some companies organise regular monthly briefing meetings between
managers and supervisors and between supervisors and the group of
employees for whom they are responsible. These meetings should on
average last about half-an-hour. Those held by production supervisors should discuss objectives and achievement in output, quality,
timekeeping, and safety, plus any other topics of current relevance.
Although production is lost when employees attend briefing meetings,
the amount of time spent at these meetings is relatively insignifi-
cant and there can be considerable benefits from the higher productivity that results from employees knowing what is expected of them.

4.3.3 Information to New Employees
When new employees are engaged, a large volume of information has to
be communicated. A representative of the personnel department can
talk on company history and products, followed by information on pay
and conditions of service. The atmosphere of induction meetings
should be informal and questions should be encouraged. The section
manager or supervisor should then show the new employee around the
section in which he will work, introducing him to his new colleagues,
and explaining details of the work he will be doing. If the company
has its own training school, it is desirable that 'green' labour
should receive a basic training there before they are put in the
factory.
  Many companies produce booklets for new employees which provide
them with helpful employment information.

4.3.4 Financial Information
The annual report and accounts of a company are not a suitable way
of passing financial information to employees. Some companies
produce an easily understood version of their financial results

designed for general distribution. The presentation of financial
information in a value-added form (see section 8.1.4) is useful in
demonstrating to employees the large amount of value added that in
many companies is absorbed by wages and salaries.

4.4 LATERAL COMMUNICATION

In large organisations, the left hand often does not know what the
right hand is doing. The production engineer may therefore be planning elaborate changes to manufacturing methods on a product that the
sales and design departments are planning to replace in the near
future. Research laboratories in different divisions of a large
company can, unknown to each other, be busy trying to solve similar
problems.

Within a product group the passage of information laterally should
be facilitated by regular meetings of the technical, production,
marketing, and financial managers.

In companies having divisional structures, those holding similar
responsibility in different divisions can hold regular meetings.
If the venues are rotated around the divisions, visitors can see
current work in progress.

At a different level, the company magazine can provide lateral
information on promotions, retirements, marriages, births, and
deaths, as well as recording the happenings of various sporting and
social clubs. These magazines are valuable in creating a family
feeling among the employees, something that is difficult to achieve
in large organisations.

4.5 UPWARD COMMUNICATION

4.5.1 The Need for Upward Communication

Although the natural flow of orders and information is downwards, it
is also important that there should be a good flow of upward communication. If management is in touch with the attitudes and reactions
of employees, the quality of management decisions is improved.

Good upward communication will enable the company to benefit from
the untapped ideas and initiative in the lower levels of the company.
It will also help to create among employees a spirit of team-work as
they know they can make their views known to management.

Many senior managers in large companies can become seriously
isolated from the attitudes and opinions of their workforce. This
isolation is often due to the protection given by secretaries and
personal assistants. Matters are made worse if senior managers
surround themselves with sycophantic staff who will communicate only
flattering news. These conditions will produce insensitive management who will sooner or later suffer a sudden and rude awakening to
the true state of affairs.

By getting out of their offices and walking around the factory,
laboratories, and offices, senior managers can talk with employees
and sense their attitudes. The occasional visit to the night-shift
or arrival on the factory floor at starting or leaving time can be
very revealing. This top-bottom contact is often resented by inadequate middle management.

4.5.2 Suggestion Schemes

These are an effective way of encouraging ideas that benefit the
company. The majority of the suggestions are those intended to
reduce manufacturing cost but ideas to improve product quality or

customer service should also be encouraged. Promising ideas are
investigated and developed and the originators of the suggestions
are appropriately rewarded. Those who submit ideas that are not
accepted are thanked and asked to try again. They should be given
a full explanation why their ideas were unacceptable.

4.5.3 Consultative Committees

These provide a formalised method of bringing the upper and lower
layers of the organisation together at regular intervals. They are
not usually needed in smaller companies since there consultation
occurs informally.

 The function of consultative committees is not to pass down management decisions but to seek the views and ideas of employees before
decisions are reached.

 In unionised companies, it is convenient if the employee representatives are union representatives as this enables matters covered
by union agreements to be included on the agenda. A wide range of
topics is discussed, including productivity, training, and promotion
policies. If union representatives are present it is possible to
discuss, at the pre-negotiation stage, such topics as wages, hours
of work, and holidays.

 The chairman should be the person responsible for the work area
covered by the committee; he should be supported by appropriate
managers and supervisors. It is advisable to constitute the committee with fewer management than employee representatives. Meetings
should take place at regular intervals, a suitable frequency being
three or four times a year, with additional meetings if required.

 A simple news-sheet should supplement the minutes of the meeting.
There should be opportunities during working hours for employee
representatives to report back verbally to their groups.

4.5.4 Safety Valves

A number of methods are available for employees to communicate
frustrations and disgruntled feelings upwards. One approach is to
have a number of complaints sheets and boxes situated in the factory
and offices. Any employee may, without stating his name, make a
written complaint on any matter affecting his work. The boxes are
emptied, the complaints considered, and any necessary action taken.
A more refined system enables any employee to make a written complaint to which the responsible manager will reply in writing. The
letter from the employee is passed through the personnel department,
ensuring that the complainant remains anonymous.

4.6 THE ORGANISATION AND CONDUCT OF MEETINGS

This section attempts to provide information on the typical semi-
formal meetings that an engineer is likely to attend in the course
of his work.

 Meetings are expensive and should be held only when there is a
real need for them. They should last no longer than is required to
achieve their objective and the number of people invited should be
kept to the minimum necessary.

4.6.1 Preparation for the Meeting

If a meeting is to be successful some preparatory work by the chairman and secretary is essential. Unless the meeting is a regular one
with a fixed attendance, the chairman will have to decide whom to

invite; from 5 to 18 people is considered a suitable number to give
reasonable discussion.

The agenda is determined by the chairman; this is an important
document as it forms the structure of the meeting. Items on the
agenda should be stated in sufficient detail for members to apprec-
iate fully what is to be discussed. Although it is difficult to
estimate the length of time each agenda item will occupy, the chairman should try to avoid a meeting that occupies more than two hours.

It is recommended that 'matters arising from the minutes' are
omitted from the agenda; if any of these require discussion, they
should be included as separate agenda items. It is also recommended
that 'any other business' should be omitted as it is unlikely that
many of those attending the meeting will have sufficient knowledge
to discuss items introduced without notice. These two recommendations are unusual and contentious but do have merit.

Several days before the meeting, the agenda and any supporting
documents should be circulated to the members by the secretary. The
minutes of any previous meeting should also have been circulated.

4.6.2 Conduct of the Meeting

The chairman's task in conducting the meeting is one that calls for
considerable tact, patience, concentration, and common sense. His
rôle is to control the meeting, allow adequate discussion, and
achieve the objective of the meeting.

At the beginning of the meeting the chairman should remind those
present of the purpose of the meeting and should obtain approval for
the way he intends to proceed. His attitude should be flexible and
he should be prepared to be persuaded by the argument of other
members. If the meeting wanders from the point he must bring it
back again. When there has been adequate discussion of an agenda
item he should close the discussion and outline the conclusions
reached. At the end of the meeting the chairman should conclude
with a brief summary of the achievements and try to send the partic-
ipants away with the feeling that their presence has been of value.

The secretary, apart from attending to the detailed arrangements
for the meeting, is available to assist the chairman during the
course of the discussion with information and advice. He also takes
notes from which he prepares the minutes.

Members attending should respect the authority of the chairman and
should address their contributions to him. They should make their
contributions to the discussion constructive, brief, and to the
point.

4.6.3 The Minutes

When the secretary has drafted the minutes they should be approved
by the chairman before they are issued. If, as a result of the
meeting, action has to be taken by some of those attending, they
should be reminded of their obligations by the secretary. A convenient way of doing this is to include an action column in the margin
of the minutes so that the initials of those required to take action
can be added against relevant items. It is essential that minutes
should be a brief, balanced, and accurate report of the meeting.

4.6.4 Speaking at Meetings

Many engineers are not naturally good speakers but most can become
reasonably competent with experience and help. Brevity and clarity

are as important in speaking as in any other form of communication.
Preparation should be made for any known or probable contribution to
a meeting. If the contribution is likely to be a major one, a list
of points in heading form should be prepared for quick reference
when speaking. The reading of a prepared text should be avoided, as
should the writing-out of a speech and committing it to memory. At
most informal meetings, the use of elaborate visual aids is unnecessary. However, if they are used, they should be well-prepared and
the speaker should ensure that he is fully conversant with their
contents, sequence, and use.

A speaker should attempt to maintain eye-contact with his audience
as this helps to keep him in touch with their reactions. If ges-
tures are used to lend emphasis to what is said, they should be used
in moderation. Should a speaker have nervous habits when speaking,
he should try to suppress them. Nothing should be said that is calculated to hurt or offend other speakers, even if what they say
appears foolish. A speaker should remain calm during debate however
strongly he feels about the subject under discussion; if someone is
rude to him he should be polite in return. Should a person wish to
speak he should first catch the chairman's eye and not blunder in
with his contribution. Finally, those attending meetings should
think before they speak and not speak at all unless they have something worthwhile to say.

Many companies organise effective speaking courses for their staff
where specialist help and advice is given on an individual basis.

4.7   REPORT-WRITING

Report-writing is frequently required of engineers since much of
their work consists of examining problems, reporting findings, and
making recommendations. Reports should be brief, accurate, and
easily understood, a requirement that demands considerable effort on
the part of the writer. It is important that the young engineer
writes good reports since they can considerably influence the esteem
in which he is held by his senior colleagues. The following material
on report-writing is not exhaustive and a number of excellent books
are available for reference.

4.7.1 Collection of Information

A report should not be started until the writer is quite clear about
its objectives; if possible the person requesting the report should
supply terms of reference. Relevant and correct information should
be collected and its source noted. A convenient way to collect data
is on cards, using one card for each piece of data.

4.7.2 Arrangement of Information

Keeping the object of the report constantly in mind, the data should
be arranged in suitable divisions and subdivisions.

A typical arrangement of a report recommending a change in manufacturing methods is shown below.

(1)   Introduction, stating recommended changes.
(2)   Cost of proposed changes.
(3)   Expected savings.
(4)   Time needed to make changes.
(5)   Difficulties likely to be encountered.
(6)   Acknowledgements for help received.
(7)   Appendix to include charts, layouts, cost studies, etc.

The arrangement of a technical paper will be different; a typical arrangement is as follows.

(1) Summary of contents, sometimes called the synopsis.
(2) Introduction.
(3) Experimental method.
(4) Results.
(5) Discussion of results.
(6) Appendix or bibliography, as appropriate.

Within each subdivision the material should be arranged in a logical order. At this stage it may become evident that additional information may have to be collected.

4.7.3 Drafting of the Report

A clear, crisp style of writing is desirable. Sections and subsections should be identified by numbers or combinations of numbers and letters. If figures are used, they should be fully labelled and placed as close as possible to the first reference to them in the text.

It is a good plan to err on the safe side where estimates have to be made of expenditure, cost savings, and time needed to make changes. It is better to be remembered as a person who meets his targets, rather than as someone who just fails to achieve them. The person reading the report will require a balanced view of the proposals being made and any limitations and difficulties that are foreseen should be stated. If other people's ideas and suggestions have been included in the recommendations, they should be acknowledged.

The report should state to whom it is addressed, those who are to receive copies, the writer's name, and the date of issue.

4.7.4 Redrafting the Report

The rough draft of the report should be reread, preferably after a few days, to check that it has fulfilled the purpose for which it was written. It is advisable to read the material aloud to check that it flows smoothly and that there are no errors in grammar or punctuation. As a final check it is a good idea to try to persuade an able and experienced colleague to read the final redraft and to comment critically.

4.8 BUSINESS LETTERS

Despite the increased use of the telephone for communication, a knowledge of letter-writing is essential for engineers. A polite, clear, and brief letter will convey a good impression of your company. A letter costs considerably more than its postage; therefore, when a letter is to be dictated to a secretary, the writer should be clear about what he wishes to say before she is called into the office.

The form of address, 'Dear Sir (Madam)' or 'Dear Mr (Miss, Mrs, Ms) Smith', ending respectively with 'Yours faithfully' or 'Yours sincerely', will suffice for almost all business letters. A heading should be used for a letter if it will shorten its contents or if your correspondent has already used one. The opening paragraph of the letter should make clear its purpose. The body of the letter should cover the subject matter in a logical manner with separate paragraphs for each topic. Paragraphs should be numbered if this is

helpful in a long letter. If the form of address is 'Dear Sir' the
first person plural should be used in the letter. It is however
possible to use the first person singular should the form of address
be 'Dear Mr'. A concluding paragraph is not normally needed unless
a summary is desirable in a long letter or a point requires re-
emphasis.

Short sentences consisting of short, easily understood words should
be used. Words and phrases that make meanings imprecise should be
avoided. Abbreviations such as ult., inst., and phrases such as
'assuring you of our best attention at all times' are archaic and
should not be used. The writer's name should be typed below his
signature; the position he holds in the organisation should also be
stated.

4.9 INFORMATION RETRIEVAL

Often an engineer has to communicate with the vast body of information that exists in his subject. The growth of written information
available to the engineer resulted in the publication of *The
Engineering Index* as early as 1892. Since then, the volume of information has grown at almost an exponential rate. A manual search
through an index such as *The Engineering Index* or *Science Abstracts*
is a daunting task. The searcher normally has time only to search
under one or two headings since each item listed under the particular
heading has to be assessed for its suitability. In all but the
simplest searches, the manual method will occupy several hours if not
days.

On-line computer searches, which have been available in the UK
since 1976, put the engineer in touch with millions of references
contained in data bases throughout the world. The computer terminals, normally situated in technical libraries and large research
departments, consist of a typewriter keyboard, a line-printer, and a
visual display unit. The terminal is connected via public telephone
lines to the local node of a network of high-speed transmission lines
and thence to the selected data base. Information is called up from
the terminal using a word, a string of words, or word fragments,
linked by logical operators such as AND, OR, and AND NOT.

The computer at the data base will, within a few seconds of receiving the enquiry, reply with the number of references available. If
there are too few references, the search may be widened; if there are
too many, the scope of the search may be reduced. At any stage the
references can be sampled on the screen of the visual display unit.
Should a print-out of the references be required, this can be obtained on the line-printer at the terminal. If a large number of references is available, it will be more economical to have them printed
out at the data base and sent on by post. It is advisable that only
a person trained in the use of the terminal should operate it, otherwise the charges for the amount of time spent on-line could be excessive.

This method of data retrieval is more flexible and usually less
expensive than manual searches. It is in general limited to references subsequent to 1970.

4.10 PUBLIC RELATIONS

4.10.1 Objectives and Organisation
It is appropriate that there should be mention of public relations

in this chapter since public relations is primarily concerned with
communication. Its objectives are to establish two-way communication
between organisations and the public in order that understanding is
established and conflicts of interest are resolved. Public relations
is not propaganda; it should not attempt to distort or falsify the
truth. In fact, the employment of public relations' techniques is
likely to demonstrate the shortcomings of an inefficient or unscrupu-
lous company rather than to improve its image.

Small and medium-sized companies are more likely to hire public
relations consultants and to use them on specific assignments. Large
companies will probably employ their own public relations staff but
call on consultants from time to time when advice or additional help
is needed.

4.10.2 The Press

The national, local, and trade press are still influential in mould-
ing public opinion. In large companies it is usual to employ a full-
time press officer who will be concerned with maintaining good
relations with the press and assisting them by answering reporters'
enquiries. The press officer also provides the press with material
for articles and news, usually by means of press releases. If there
is information of major importance to be released, a press conference
is frequently arranged at which the chairman, or some other senior
member of the company, makes the announcement and answers questions.

At technical conferences, arrangements should be made for journal-
ists to sit near to the platform. If the proceedings are highly
technical, daily briefing sessions may be needed to help the reporters to understand what has been presented.

4.10.3 Radio and Television

Neither the BBC, in its domestic programmes, nor independent television and radio allow commercial organisations to influence the
contents of its broadcasts. Indirect publicity may, however, be
obtained through the inclusion of news items as part of news bulle-
tins or features programmes. The overseas service of the BBC is
interested in promoting the interests of British industry abroad and
will mention companies by name provided that their achievements are
news-worthy.

4.10.4 Other Methods of Communication

Advertising, sales conventions, exhibitions, and trade fairs all
offer opportunities for the public relations department to promote
better understanding of the company. These are also the province of
the marketing department and both departments will need to work
closely together to achieve the best results.

There are special groups with whom the company will wish to maintain good relations; these include shareholders, distributors, and
Members of Parliament. For instance, some large companies arrange
a number of regional shareholders' meetings where films are shown of
the firm's activities, questions are answered, and light refreshments
served. A loyal and well-informed body of shareholders will act as
ambassadors for the company, they could be helpfully loyal in the
event of a take-over battle, and will be more inclined to subscribe
if the company decides to issue new capital.

When the company wishes to communicate with influential individuals
or groups, it is usual for the public relations department to arrange

the meeting over a carefully planned luncheon or dinner party. New
friends should be won and it is thought equally important to maintain
contact with old friends who can help the company. Although many
feel strongly that the use of hospitality is a form of bribery, there
are others who maintain that it does nothing more than to provide an
opportunity to talk in a relaxed atmosphere. Anyone who considers
public relations to be merely an excuse in wining and dining is
advised to read the book by S. Black, listed below.

FURTHER READING

Black, S., *Practical Public Relations* (Pitman, London, 1976).
Chappell, R.T. and Read, W.L., *Business Communications* (Macdonald
and Evans, Plymouth, 1979).
Garnett, J., The Manager Responsibility for Communication, *Notes for
Managers No. 2* (Industrial Society, London).
Gode, W., Chairmanship and Discussion Leading, *Notes for Managers No.
26* (Industrial Society, London).
Kenny, P., *Public Speaking for Scientists and Engineers* (Adam Hilger,
Bristol, 1982).
Mitchell, J., *How to Write Reports* (Collins, Glasgow, 1974).

5 Marketing

5.1 THE MARKET-ORIENTATED COMPANY

Marketing is concerned with matching company capabilities to customer requirements. A production-orientated company that considers that its sales personnel are employed just to sell what its engineers decide to design and manufacture is unlikely to succeed. In the past, many producers of industrial goods have lagged behind those in the consumer market in their sensitivity to customer needs.

A company should identify its market carefully and should continuously be aware of alternative ways of satisfying customer requirements. Had the railway companies in the United States appreciated that they were in the transport business, rather than just in railway transport, they would have avoided near-extinction by the domestic airlines.

The marketing director is a key figure in ensuring that the company is market orientated. He acts as the interpreter between the market and the company and should work in co-operation with his colleagues from the other major departments. Both he and the engineering director are concerned with the performance, appearance, quality, and reliability of the company's products. He is jointly interested with the production director that customers are supplied promptly from stock or, in the case of special orders, by the promised delivery dates. Further, the marketing director and the financial director are together interested in how company profit is affected by sales volume. A graph indicating how the total unit cost of a product can reduce with volume of sales is shown in figure 5.1. Also shown in the figure is the profit or loss which is the difference between selling price and the unit cost.

5.2 THE MARKET

Consumer goods, particularly necessities, tend to experience a consistent demand. The purchase of capital goods is, however, intermittent and frequently postponed in times of recession.

Industrial goods can either have vertical demand, when sold to a single market, or horizontal demand, if sold to a number of markets. A company supplying back axles to the motor industry experiences a vertical demand for its products, whereas a company manufacturing fork-lift trucks is operating in a market with horizontal demand. Owing to their greater diversification, markets with horizontal demand generally are more stable than those where the demand is vertical.

In many industrial and consumer markets, an analysis of sales by customer shows that the bulk of the business is done with relatively few customers; typically 80 per cent of the business comes from 20 per cent of the customers. This Pareto-type distribution of sales is shown in figure 5.2.

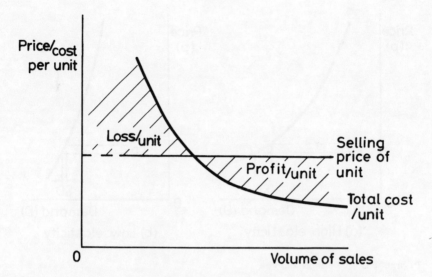

Figure 5.1   Effect of sales volume on profit per unit of sales

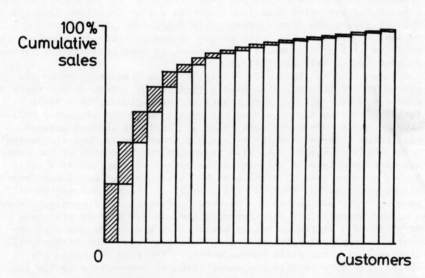

Figure 5.2   Distribution of cumulative sales to customers (actual
             sales are cross-hatched)

5.2.1   Factors Affecting Demand
The way in which price affects demand can be shown by demand curves.
In general, consumer goods tend to follow an elastic demand curve,
whereas industrial goods show an inelastic demand. It will be seen
from figure 5.3a and b that a small decrease in price produces a
considerable increase in demand when the demand curve exhibits high

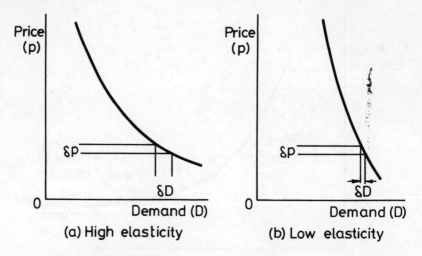

Figure 5.3   Demand curves

elasticity. Reverse inelastic demand can be experienced for some
products; here price reductions keep buyers out of the market in the
hope of further price reductions whereas a rise in prices produces a
rush of buyers who fear further price increases. Apart from the
company's pricing policy, the amount of money it spends promoting its
products will affect demand. The amount of money competitors spend
on advertising will also affect sales.

Fluctuations in the general level of trade, caused by booms and
slumps in the economy, will have a significant effect on the sales of
most products. Running down stocks at the beginning of a slump or
building them up during a boom can have a sharp but temporary influence on sales. Some products, such as gas fires and sun glasses,
experience a strong regular seasonal fluctuation in domestic demand.

The level of competition will vary with the structure of the market
and is generally more intense if the market is supplied by a large
number of companies of similar size. A complete monopoly will remove
all competition but this is most unusual in the private sector of the
economy. The Monopolies and Mergers Commission can investigate where
a single organisation supplies more than 25 per cent of the goods and
services in a stated market or where there are agreements that prevent the supply of particular goods and services. Competition will
also vary with the state of the market. The more depressed the
market, the more intense the competition as companies scramble for
survival.

5.2.2 Market Share
Market share is the proportion of the total sales in a given market
held by a particular company. In general, companies with a large
market share are more profitable than those with a small share, since
the economies of scale work in their favour (see section 2.4.2).

It is therefore to be expected that marketing directors normally
will strive to increase their company's share of the market. However,
it may be decided to consolidate if either it would be too expensive
to capture more of the market or a larger market share may risk an

unwanted investigation by the Monopolies and Mergers Commission.
Thirdly, it may be decided that the company should reduce its share
in a particular market by freezing product development and under-
spending on promotion. In this way, higher profits can be reaped in
the short term prior to complete withdrawal from the market.

5.3 SALES-FORECASTING

Sales-forecasting is a hazardous operation which despite its inaccu-
racies must be attempted. Long-term forecasts are required for strategic planning, medium-term forecasts are needed for budgetary
purposes, and short-term forecasts are used for production-planning.
   Forecasting techniques vary from the simple to the complex. Accuracy does not necessarily increase with the complexity of the technique used but accuracy can be improved by combining the results
obtained by two or more different methods. Techniques fall into
three broad categories: those based on opinion, those based on the
analysis of historical data, and those in which sales are predicted
from one or more variables that can be linked with product demand.

5.3.1 Forecasts Based on Opinion

The opinion of sales representatives and distributors can be used as
a basis for forecasts. Sales representatives should be aware of
sales prospects in their territories, the strength of the competition,
and the attitudes of their customers. Although sales staff will
normally complete surveys carefully, their opinions tend to be over-
optimistic or over-pessimistic, depending on market conditions. It
may be difficult to persuade distributors to complete surveys but
their surveys provide a more balanced view of the market since they
are selling a range of goods.
   The level of demand for new products is always difficult to gauge
but an indication of eventual sales can be obtained by a trial marketing of the product on a limited scale.

5.3.2 Forecasts Based on Historical Data

This is probably the most widely used method of sales-forecasting.
A simple method is the 'eyeball approach' in which sales are plotted
against time and the graph is projected forward by eye (see figure
5.4). Mathematical methods of establishing a trend-line include
least-squares regression, for points lying about a straight line, and
curve-fitting. Seasonal variations in demand can be eliminated by
plotting annual sales.
   Moving-averages can be used to smooth random fluctuations in sales.
By weighting the moving-average, greater emphasis can be given to
more recent sales levels. A systematic method of weighting, called
exponential-smoothing, weights the data exponentially. Exponential-
smoothing can be easily handled by a computer and is widely used for
ordering parts. A more detailed explanation of the above methods of
projecting historical data can be found in *The Management of
Manufacturing Systems* by J.D. Radford and D.B. Richardson. The
projection of historical data assumes that existing trends will be
maintained; this is not always true but it should be possible to
superimpose corrections once the trend-line has been established.

5.3.3 The Use of Correlation

A third method attempts to find one or more statistics that correlate
with past sales and from which future sales can be forecast. Some-

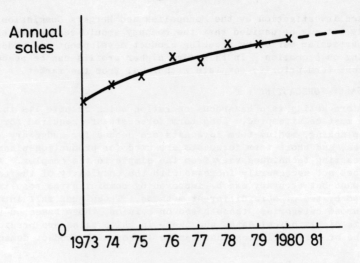

Figure 5.4   Extension of sales graph

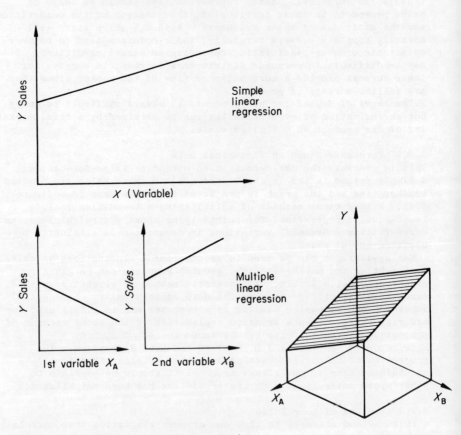

Figure 5.5   Types of linear regression

times the better-known economic statistics, such as consumers' expenditure on goods and services, will correlate with product sales. On
other occasions the statistic will be much more specific, such as the
new registrations of agricultural tractors. If sales are related to
a single variable, the relationship is one of simple regression: if
they are related to two or more variables, the relationship will be
one of multiple regression. Relationships can be linear or curved.
The examples shown in figure 5.5 are for simple linear regression and
for multiple linear regression.

5.4 BUYER MOTIVATION

It is useful to consider the factors that influence buyers' decisions.
There are the obvious economic factors of price, delivery, and quality. In addition, the preferences of the buyer can intrude and are
influenced by social, political, ethnic, and cultural attitudes.
   When an important non-routine purchasing decision has to be made
it is possible that several people will be involved, some of whom
will occupy senior positions in the company. One of these people has
been identified as the gatekeeper; this person collects preliminary
data and sends out enquiries to suppliers. The user can often have
a strong voice in the choice of supplier. Other influencers of
opinion will include technical journals, the trade press, sales
representatives, and the opinions of other companies. The buyer himself may be the decider. On other occasions a technical expert, such
as a production engineer, will be the person who makes the final
decision on where to purchase.
   Major buying decisions often involve considerable risk and cause
considerable anxiety. Good after-sales service and technical back-
up is helpful in setting minds at rest. Some manufacturers, realis-
ing post-purchase concern, even produce advertisements congratulating
purchasers on their sound choice in buying their products.

5.5 MARKETING RESEARCH

5.5.1 The Need for Marketing Research

Decisions of major importance to the future shape and prosperity of
the company have to be based on recommendations from the marketing
department. To make the right decisions on product policy and marketing, appropriate questions must be asked; marketing research helps
to answer these questions correctly. It is usual in most companies
to subcontract marketing research projects to specialist organisations.

5.5.2 Marketing Research Procedure

The company should carefully brief those undertaking the marketing
research and agree with them its feasibility, accuracy, and time-
scale.
   Data can be obtained either by reactive or non-reactive research.
Reactive research involves obtaining information by the response of
those questioned or from those taking part in an experiment. The
questionnaire is the most popular method of gathering marketing
research information. The questions can be asked either by post or
by an interviewer. Considerable skill is needed in the design of the
questionnaire and in the selection of the sample to be taken.
   Well-designed experiments can provide useful marketing information.
For instance, a chain store could ascertain whether greater sales

will result from special offers for a limited period or from smaller
permanent price reductions. Two similar sales outlets could be
selected for the test and, during a trial period, one could make
special offers, while the other makes permanent price reductions.

Non-reactive marketing research can either involve observation or
the analysis of existing data. If observation is employed, the
reactions of people to specific buying or promotional situations are
recorded and interpreted. When existing data are used this is referred to as desk research; data sources include information published
by international agencies, the government and trade associations, or
perhaps unpublished company information. Desk research is a good
starting point for any marketing research project since it is a waste
of money to collect data that are already available.

Random sampling can be used in the consumer market. In the industrial market some type of quota sampling is often used since the
market is made up of companies whose sizes are skewed.

A report based on the information gathered is submitted. Most
companies welcome an interpretation rather than a bare statement of
the findings. However, there is some feeling that those engaged in
marketing research should deal only in facts.

5.6 MARKETING PLANNING

Marketing planning is a starting point for making corporate decisions. Planning in competitive markets against a background of
uncertain economic conditions means that original plans frequently
have to be updated and revised. Despite these uncertainties, objectives must be set and plans made to achieve the objectives.

5.6.1 Setting Objectives

As a prerequisite of marketing planning a statement should be made
by the directors to define the market in which the company intends
to operate. A company manufacturing expensive pens may wish to
distinguish itself from makers of inexpensive ball-point pens by
stating that it is in the high-quality gift market rather than in
the writing-instrument market. Markets should be surveyed for
opportunities and the activities of competitors examined. The
volume, classification, and geographic location of future markets
should be carefully considered.

It is probable that objectives will be divided into short- and
long-term ones. Long-term objectives are concerned with strategic
planning and look forward over a period of several years. The minimum period covered is often determined by the lead-time needed to
plan and build additional production capacity or to design and
develop new products. Short-term objectives frequently have a
horizon of one year or less and are concerned with tactical decisions.

5.6.2 Strategic Planning

Long-term marketing objectives are pursued by trying to select the
appropriate marketing mix. The factors that make up this mix are
product policy, price policy, distribution policy, and marketing
communication policy. Marketing communication policy comprises
advertising, selling, and other promotional activities.

The products should be considered in relation to customer needs,
competitors' activities, and the design and manufacturing capabilities of the factory.

Pricing is a crucial factor in the achievement of objectives. Price-acceptability studies should be conducted and competitors' prices examined. It is often necessary to fix the price and then to design a product that can be sold at that price. Determining price by adding a fixed percentage to the total cost of the product lacks flexibility. For instance, it may be desirable to sell at a loss for a limited period to penetrate a particular market. Further, it may be sound policy in times of depression to accept some orders at selling prices that cover only material, labour, and variable over-head costs but make no contribution towards covering fixed overhead costs.

Advertising, which is discussed in greater detail in section 5.7, is usually handled by an agency. It is important that the agency's work is fully integrated into the strategic marketing plan.

Selling methods should be compatible with the market and will vary considerably between consumer and industrial markets. Industrial salesmen usually require good technical knowledge and back-up since customers are increasingly asking for their problems to be solved rather than themselves deciding which catalogue items to order.

5.6.3 Tactical Planning
Strategic plans are more likely to be achieved if they are broken down into short-term tactical tasks and objectives. Sectional managers are often given their short-term broad objectives then asked to split these into individual tasks. For instance, the area sales managers will break down their sales targets into quotas for individual salesmen. Performance against targets should be regu-larly monitored and differences investigated.

5.7 COMMUNICATIONS WITH CUSTOMERS

5.7.1 The Communications Mix
Marketing information can be given to customers in two ways; in-directly through advertising or directly through personal selling by sales representatives.

There are many ways of reaching customers by advertisements. The more important are newspapers, magazines, journals, television, direct mail, and product-packaging.

Personal selling can also take a variety of forms; sales repre-sentatives can call regularly on established customers or they can visit prospective customers with the object of obtaining new business. The call can be made cold, although this is unusual when selling industrial goods. Exhibitions and trade fairs provide opportunities for sales staff to demonstrate products and to meet old and new customers. They also provide an opportunity to assess the strength of the competition.

5.7.2 Objectives of Advertising
Four major objectives can be identified; the emphasis on each will vary with marketing strategy. These objectives are to create a favourable image of the firm, to inform the public about products or services offered, to influence choice, and to reinforce loyalty.

Advertising is a significant item of expenditure and can rise to about 30 per cent of the sales income of some manufacturers of consumer goods.

5.7.3 Advertising Agencies

Most companies subcontract their advertising to agencies. An agency
should be carefully chosen and given a clear explanation of the rôle
of advertising in the company's marketing plan. The agency designs
advertising material and selects where it should be used. Material
such as television advertisements is often given pre-release checks
by sample groups who assess its effectiveness.

Advertising agencies will organise stands at exhibitions and will
arrange for advertising films to be made if required.

5.8 EXPORTING

Britain's share of overseas trade has fallen sharply in the twentieth century. The export of manufactured goods is, however, essential if we are to pay for imports of food, materials for industry,
and foreign manufactured goods.

Although successive governments have encouraged exports, it is the
urge of individual companies to seek profits overseas that creates
export markets.

5.8.1 Levels of Export Involvement

The commitment of a company to export may have developed from a
chance enquiry from overseas rather than from a carefully planned
attack on a particular market.

Several stages can be identified in the development of a company
into a fully international business. Initially the company may have
seen and seized a chance opportunity to sell abroad. The next stage
could be a commitment to continue selling abroad into selected
markets, using factory output that is surplus to that needed to
supply the home market. This could be followed by developing products specifically for the export market and the recognition that
exporting is a significant part of the company's activities. Finally
the company may become multi-national, manufacturing and selling
throughout the world and shifting investment to where it is expected
to earn the highest rate of return.

5.9 EXPORT PRODUCTION POLICIES

5.9.1 Home Manufacture

However large the export market, it is still possible to supply its
demands from home factories. This policy gives the company greater
control over production and quality although adding to distribution
costs.

5.9.2 Building Factories Overseas

The reasons for building factories abroad include lower production
costs, avoidance of import duties, and meeting the wishes of foreign
governments. Governments of foreign countries often require a total
or partial local involvement in product manufacture. Investment in
overseas factories should be made only after a thorough assessment of
the financial and political risks.

Joint ventures with local companies offer an alternative to providing all of the capital and expertise needed to set up overseas manufacturing facilities. It is difficult to avoid friction in joint
ventures and frequently they end in a take-over by one of the parties
or in an agreement to separate.

5.9.3 Licensing

This arrangement enables a local company to produce a product or to

use a process protected by patents. The licensing company usually
receives a down payment plus royalties based on output. There is no
investment required nor any financial risk to the company granting the
licence. Expert legal and commercial advice is necessary in the
preparation of licensing agreements.

5.10 EXPORT SELLING POLICIES

5.10.1 Direct Marketing
This is well suited to large companies and provides close control.
Smaller companies, which often find direct selling too costly and do
not wish to use agents, may engage export merchants who will provide
full marketing facilities.

5.10.2 Agents
Selling can be let out to agents who import goods at agreed prices
from the manufacturer. The agents stock, sell, and distribute the
goods in a defined territory.

Another form of agent is the commission agent who is paid an agreed
rate of commission on the value of orders sent to the company. Com-·
mission agents usually deal in expensive capital equipment for which
demand is intermittent.

Considerable local investigation is desirable before an agent is
appointed and the agency agreement should be carefully drafted. Some
agents will deal solely with the products of a single company whereas
others deal with a range of products but not normally competitive
products.

5.10.3 Barter
This type of trading is often used with communist countries and with
underdeveloped countries that are short of hard currencies. It is
often less profitable than normal trading and involves the exporter
receiving payment in the form of goods or commodities which he then
has to sell. Companies may find barter useful in disposing of goods
surplus to normal requirements; it also enables trading to take place
in markets that otherwise would be closed.

5.11 PRODUCT INNOVATION

A policy of sound product innovation is central to the prosperity of
a manufacturing company. A complacent company that fails to develop
new products will eventually find itself in difficulties and one that
develops unsuccessful products can rapidly make large losses. Companies seeking growth should develop new products at a faster rate
than is needed to replace the ageing ones in their present portfolio.
Although new product policy will depend on advice from the marketing
director, it also should be the concern of the whole management team,
particularly the engineering and the production directors.

When considering new product policy, it is helpful to assess the
current market position of the company's products. This can be done
by constructing a matrix in which the competitive positions of the
company's products are shown against the market prospects for the
type of product. In figure 5.6 the position of three products sold
by the company is shown relative to their two leading competitors in
each market. In this instance it would not appear to be worth while
developing a new product to replace A3, whereas successors to A1 and
A2 should be developed when appropriate.

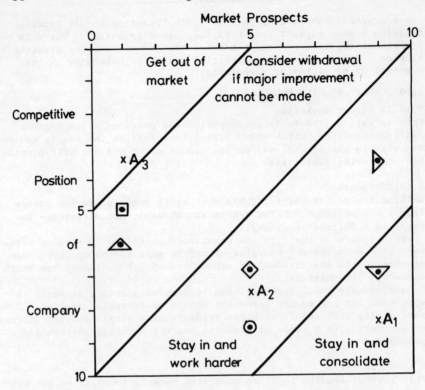

Figure 5.6 A product matrix

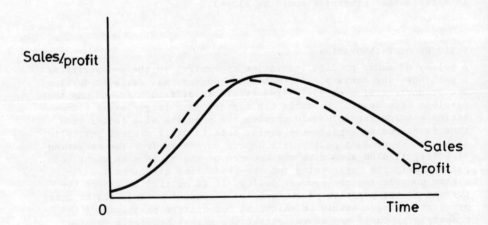

Figure 5.7 Product life cycle

5.11.1 Product Life Cycles

Companies must be aware of the life cycles of their products. A typical life cycle curve for a product is shown in figure 5.7.

From past experience a company will know the life span of previous similar products; other useful guides to the position of a product in its life cycle are its sales and profitability. New products should be planned to take over as products slip from middle- into old-age. A company should aim to have a good spread of young, middle-aged, and elderly products. Product life cycles cannot be considered fixed and will be affected by technical developments and competitive activity. A company must be ready to mount a crash new product-development programme to replace a prematurely aged product.

5.11.2 New Product Development

There are a number of ways in which a company can develop new products.

Improving and Adapting Existing Products

Improving a present design can be relatively inexpensive and should prolong its profitable life.

The technique of value analysis can be used to consider ways of improving the value of the product to the customer; it is described in section 5.11.4. Design consultants can be employed to 'face-lift' the appearance of the product. Improvements to products can be introduced piecemeal or consolidated with other innovations to create new models.

Sometimes a product sold only in the home market can be modified for export sale. Also a manufacturer may find that his product can be adapted for uses that are different from those for which it was originally intended.

Seeking New Products from External Sources

New products that have been developed elsewhere can be made either under licence or as joint ventures: a third possibility is to purchase them outright from the inventor. The Rank Organisation's agreement with Xerox to manufacture their dry copier converted Rank into one of the most successful British companies of the 1960s.

Another method of acquiring new products is to take over another company that has recently developed a promising product.

Inventing a New Product

Most companies are never involved with the invention of a major new product. Research and development departments are expensive and cannot be guaranteed to produce a flow of worthwhile inventions. Often the discovery of new products, for example stainless steel, are stumbled upon when looking for something else. Major new inventions do not always come from research and development departments; the Xerox copier was invented by a patents lawyer, automatic telephone dialling by an undertaker, and Kodachrome by two musicians.

5.11.3 Introduction of New Products

There is little point in designing a new product unless it can be produced and sold with minimum delay. In consequence, when a new product is to be launched its design, manufacture, and marketing should be considered as an integrated operation.

The main steps in the introduction of an engineering product to be sold in quantity are outlined below. It has been assumed that the initial development work on the product is complete and that the financial expenditure has been authorised.

(1)   Design product.
(2)   Make and test prototype of design. Modify if necessary and retest.
(3)   Apply value engineering to the design.
(4)   Determine where parts will be made.
(5)   Release drawings and specifications.
(6)   Obtain bought-out parts.
(7)   Plan method of production of factory-made parts.
(8)   Design and manufacture production-tooling.
(9)   Obtain manufacturing equipment, if not available.
(10) Obtain material.
(11) Promote the new product.
(12) Recruit and train necessary labour.
(13) Manufacture parts.
(14) Assemble and test production prototypes. Modify if necessary and retest.
(15) Commence deliveries to warehouse.
(16) Commence distribution to customers.

Some of these operations are end-on. For instance, tooling cannot be designed until a decision has been made on the method of production. Other operations may overlap; an example of overlapping operations is where tooling is being designed and manufactured concurrently with the procurement of bought-out parts and materials.

To ensure that the whole project is completed within an appropriate time, the duration of each activity should be determined. Activities must then be combined in a logical sequence to provide a total project time. An appropriate way of doing this is to use a network diagram. The construction of network diagrams is explained in section 10.4. If the original project time is unsatisfactory it can be changed by re-allocating resources. A typical network diagram for the introduction of a new product is shown in figure 5.8a. Single activities in this network can be expanded into greater detail; this has been done for the first activity, 'Design', and is shown in figure 5.8b. It will be seen that the activities in this detailed network have been drawn against a time-base, making it suitable for progressing purposes. Non-critical activities are shown to occur as late as possible with their tail slacks shown by broken lines; dummy activities are given by broken lines along their entire length. A more easily understood method of displaying the information on a network diagram is to use a Gantt chart. The work scheduled by the network in figure 5.8b is shown in Gantt-chart form in figure 5.8c (see page 84).

5.11.4 Value Engineering

Value engineering is an organised way of challenging unnecessary cost in design. It is also concerned with providing increased value to the user, therefore the redesign does not always result in reduced manufacturing costs. Utility may be increased or there may be some specialised advantage gained, such as weight-saving in aircraft. The ideas used in value engineering are not new; they were first codified

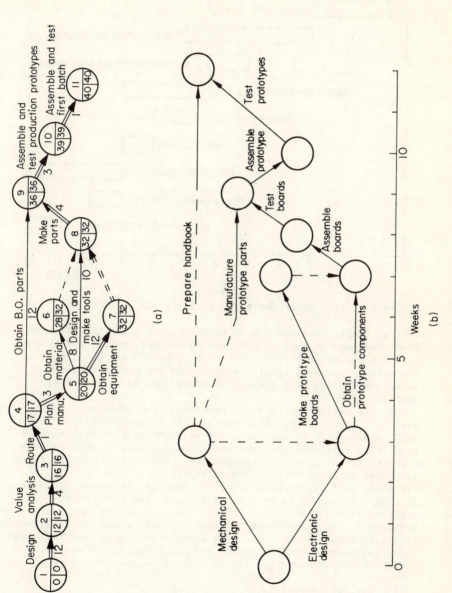

Figure 5.8 (above and on page 84) (a) Network for the introduction of a new product (b) Progressing network for activity 1-2

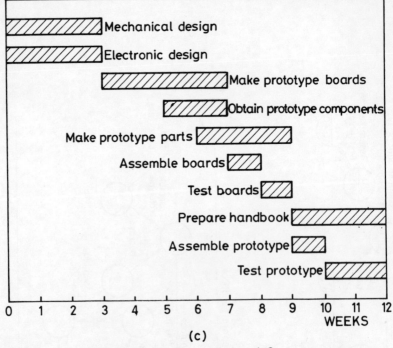

Figure 5.8   (c)   Gantt chart for activity 1-2

in a generally applicable form in the United States in the late 1940s.
Since many new designs do not provide the best value, and because
most designs are not fully updated to take account of changes in
technology, there is usually ample scope for design improvement.
Value engineering is a simple technique that has received wide industrial acceptance.

The term 'value analysis' is sometimes used synonymously with value
engineering. However, value engineering is specifically used to
imply application of the technique between the prototype and the
production-tooling stages whereas value analysis is applied to a
product that is already in production. Specialists in this field
often are referred to as value engineers.

The basis of value engineering is, however, teamwork; the membership of, the teams will vary but a typical team could consist of
representatives from design, production, accounting, purchasing, and
work study. In some instances value analysis teams are constituted
on a full-time basis but the more usual arrangement is a full-time
value engineer co-ordinating the work of a part-time team.

Six stages can be identified in a value engineering study

(1)   Setting objectives.
(2)   Collecting information on the cost and function of the
existing design and its component parts.
(3)   Speculating on ways of improving value.
(4)   Investigating proposals.
(5)   Recommending design changes.
(6)   Implementing design changes.

Setting Objectives

It is desirable that broad objectives be set in consultation with
senior management; in this way the team does not waste time on
unacceptable proposals. For instance, if a company is manufacturing
electrically operated water heaters, is it going to be interested in
other water-heating methods? The team members should acquaint themselves with any particular design limitations, such as size or operating environment. They may also have to be informed of a time limit
by which the value-engineered design should be ready.

Collection of Information

This preparatory work should be organised by the value engineer, who
provides an analysis of the design that indicates the function, cost,
and source of each part. Information is also provided on the expected rate of production and on the sales-life of the product. If a
part is made in the factory, details of the division of costs between
material, labour, and overheads should be available. A graphic
method of indicating to the team the concentration of cost in a
relatively small proportion of parts is to draw a Pareto distribution.
This is constructed by arranging the parts used in the product in
descending order of cost and then plotting the cumulative cost
against the number of parts, starting with the most expensive part.

Speculation

At this stage, the team generates the value-improvement ideas. It is
important that there should be a sympathetic chairman who can produce
an atmosphere where seniority is forgotten thus encouraging a free
and frank exchange of ideas. By bringing together people from different departments to generate ideas, surprisingly better results are
achieved than if each member works in isolation. Sometimes suppliers
may be invited to join the team; on other occasions customers can be
asked to contribute. In engineering products some of the likely
areas of value improvement are as follows.

(1) *Function* Are all the functions performed by the design essential? Are the essential functions performed satisfactorily?

(2) *Number of parts* Could the number of parts be reduced?

(3) *Processes* Are the production processes specified the most
economical, taking account of the quantities required?

(4) *Material* Could less or cheaper material be used? Could
material wastage be reduced?

(5) *Quality* Is it possible to reduce quality levels? Can quality
be increased thereby reducing assembly times and the rectification
costs?

(6) *Standardisation* Could standard components or materials be
used?

(7) *Suppliers* Are there other suppliers who could supply more
cheaply while maintaining adequate standards of quality and delivery?

All ideas generated at this stage are noted; value judgements should
not be passed on ideas since they are likely to restrict their flow.

Investigation

Here ideas from the previous stage are sifted and the most promising
costed. This can be a time-consuming task since both customers and
suppliers may have to be consulted.

Recommendation

At this stage the selected ideas are considered by the team and a decision made on those to be recommended for incorporation in the improved design. Prototypes of the redesign may have to be made and tested.

Implementation

After management approval has been obtained for the design changes, any necessary tooling should be manufactured and parts ordered. An implementation date will have to be determined, taking account of stocks of superseded parts, as well as of the procurement time for new parts.

For value engineering to be successful, it should receive backing from management at the highest level. This will assist in breaking down departmental barriers and reducing delays particularly at the implementation stage. It should be appreciated that meetings of value engineering teams are expensive and they should not be allowed to degenerate into a pleasant escape from the stress and rush of the factory. This risk can be minimised if the chairman has to report progress to senior management at regular intervals.

FURTHER READING

Berridge, A.E., *Product Innovation and Development* (Business Books, London, 1977).

Blakstad, M., *The Risk Business, Industry and the Designer* (Design Council, London, 1979).

Branch, A.E., *The Elements of Export Practice* (Chapman & Hall, London, 1979).

Broadbent, S., *Spending Advertising Money* (Business Books, London, 1979).

Carson, J.W. and Rickards, T., *Industrial Product Development* (The Gower Press, Farnborough, 1979).

Chisnell, P.M., *Effective Industrial Marketing* (Longmans, Harlow, 1977).

Chisnell, P.M., *Marketing Research* (McGraw-Hill (UK), Maidenhead, 1981).

Radford, J.D. and Richardson, D.B., *The Management of Manufacturing Systems* (Macmillan, London, 1977), pp.12-25.

Roberts, J.C.H., *Value Analysis, ABC of* — (Modern Management Techniques, Southport, 1967).

6 The Engineer and the Law

This chapter has been designed to provide a brief review of some of
the legislation relevant to an engineer in his work; those requiring
more detailed and legally precise information are referred to the
bibliography at the end of this chapter. The type of work done by an
engineer will influence which aspects of industrial law are likely to
be most relevant to him; for instance a design engineer is likely to
be interested in patent law whereas a factory manager will be more
concerned with legislation relative to health and safety at work.

6.1 SETTLEMENT OF DISPUTES

English law can be divided into criminal and civil law. The criminal
law is concerned with offences in which the wrongdoer is prosecuted
by the state. Examples of such offences are those contravening the
statutory provisions of the Health and Safety at Work Act. Civil law
regulates the relationships between individuals. Conflicting claims
involving individuals and companies can be settled by the civil
courts or by industrial tribunals with the police taking no part in
the proceedings.

Arbitration is another way of settling commercial and industrial
disputes. Both parties must agree to this course of action and be
prepared to accept the arbitrator's award. The arbitrator or arbi-
trators are neutral persons acceptable to both sides. Arbitration is
a quicker and cheaper way of settling commercial disputes than refer-
ring them to the courts.

In trade disputes, conciliation may be used to seek a solution;
here one party asks a conciliator to try to achieve a solution. Conciliation can be attempted without invitation from either party to
the dispute; this is sometimes done to try to settle serious industrial disputes.

6.2 TORT

Any person is protected under the general law against the torts of
others; these wrongs include nuisance, negligence, libel, slander,
trespass, and assault. An employer is responsible for torts committed
by his employees in the course of their employment; the employer
escapes liability only if the tort results from something that he has
expressly forbidden the employee to do. This responsibility of the
employer for the torts of his employees is called vicarious liability.
An employer is not, however, generally responsible for the criminal
liabilities of his employees.

6.2.1 Nuisance
This occurs when there is unreasonable interference with individuals
to use or enjoy their land. Nuisance could arise if the factory emits

an unreasonable volume of noise, unpleasant smells, or is a source of vermin. The complainant can apply to the court for damages or for an injunction preventing the nuisance. The police are responsible for dealing with public, as opposed to private, nuisances.

6.2.2 Negligence
Negligence is proved if it can be shown that unintended damage has resulted from a failure in the legal duty of care. This duty is owed to those one knows or should have known would be likely to be affected by one's conduct. The test of whether care has been exercised is what a reasonable person would have done to avoid injury. If one is professionally qualified, one's action will be compared with that of someone from that profession acting reasonably.

Whatever safety precautions are taken by employers, certain work remains inherently dangerous; for example, working with explosives or deep-sea diving. Those who undertake this type of work and are injured have no claim against their employers if their injuries result from the inherent occupational dangers that they have accepted.

Visitors to a factory are 'owed a common duty of care' and must be properly warned of potential dangers. This care need not be shown to trespassers. However, if people trespass to the knowledge of the owner, for instance local youths regularly playing football in the factory yard, they will be treated by the law as visitors and not trespassers.

6.2.3 Passing-off
The tort of passing-off is committed if a company intentionally or otherwise gives a reasonable person the impression that its goods or services are connected with another company. This enables a company that has already established a market for its products to sue another that causes the public to think that its products are those of the established company.

6.3 LABOUR LAW
British labour law reflects the growth of trade union power and the expectations of individual workers for continuing improvements in the terms and conditions of work. When most people worked in agriculture, the law of contract was sufficient to regulate the master and servant relationship. In today's welfare state and complex industrial society there are extensive statutory requirements, in addition to the law of contract, that the employer must observe in his relationship with his employees.

6.3.1 Law of Contract
Although most labour law is recently enacted statute law, the law of contract is still of interest in relation to contracts of employment. The six requirements of a valid contract are dealt with below.

Intention
The parties to a contract must have the intention that the contract will be binding. This is assumed to be so with contracts of employment.

Agreement
There must be an agreement based on an offer and an acceptance. In
practice, an employee is normally offered employment based on standard terms which he is free to accept or reject.

Consideration
The consideration is the agreed wages offered by the employer in
return for the employee's agreement to work according to the contract. The amount of wages offered were at one time freely agreed
between the employer and employee; today they must satisfy specific
requirements. For instance, the wages in certain occupations should
correspond to those that have been agreed nationally between organisations representing employers and employees.

Capacity
Employees below the age of 18 have limited capacity to enter into
contracts; however, contracts of employment for those under 18 are
valid provided that the agreement in general is substantially for the
employee's benefit.

Consent
The terms of the contract must be freely accepted by both employer
and employee.

Legality
The contract must not be illegal; for instance, there must be nothing
in it designed to defraud the income tax authorities.

6.3.2 Duties of Employers
The main common law and statutory duties of employers are summarised
below.

   (1) To pay contractually agreed wages.
   (2) To observe provisions relating to hours of work, holidays, and
sick pay.
   (3) To permit employees time off from work for certain public
duties.
   (4) To indemnify employees for expenses arising from their work.
   (5) To provide references although the provision of character
references is not obligatory.
   (6) To insure employees against injuries at work.

6.3.3 Duties of Employees
Employees also have duties to their employers; the main duties are
as follows.

   (1) To be present and to be willing to work.
   (2) To use reasonable care and skill.
   (3) To obey lawful orders.
   (4) To take reasonable care of the employer's property.
   (5) To act in good faith; for instance, they should be honest and
they should not disclose confidential information.

6.4 AGENCIES FOR SETTLING LABOUR DISPUTES
When labour law was governed by the law of contract, disputes were
brought before the civil courts. Owing to the vast growth of employ-

ment legislation in the past 20 years, additional machinery has been
introduced to hear the greatly increased number of disputes.

Most disputes between employers and employees are resolved either
within the factory or by a disputes procedure agreed between employers and unions. If it is necessary to go outside normal procedures
then one of two bodies will probably be involved in the settlement.
The Advisory, Conciliation, and Arbitration Service (ACAS) is likely
to be involved in trade disputes or, if the dispute relates to dismissal, redundancy, or discrimination, the dispute could come before
an industrial tribunal.

6.4.1 The Advisory, Conciliation, and Arbitration Service

This is a statutory body set up to promote better industrial relations. At the request of either party to a trade dispute, ACAS will
attempt to conciliate and settle the dispute; it may also attempt to
conciliate without an invitation. Should both parties agree to
arbitration, then ACAS will appoint an arbitrator.

ACAS can carry out inquiries in the field of industrial relations
and propose codes of practice designed to improve industrial relations and thereby avoid disputes.

6.4.2 Industrial Tribunals

These were set up to settle disputes under the Industrial Training
Act of 1964; since then their jurisdiction has been greatly increased
to deal with complaints under a large number of labour laws.

Before a formal hearing of the tribunal takes place, an attempt is
usually made to settle the complaint by conciliation.

Tribunals are organised on a regional basis and have a legally
qualified chairman assisted by two lay members; one is nominated by
an employers' association and the other by a trade union.

Appeals on points of law concerning decisions of industrial tribunals are heard by the Employment Appeals Tribunal. The president of
this tribunal is a judge who is assisted by lay members having
special knowledge or experience of industrial relations.

6.5 STATUTE LABOUR LAW

Statute labour law is to be found in a considerable number of Acts.
The position is somewhat confused owing to consolidation and amend-
ment. For instance, the Contracts of Employment Act, 1972 was combined with parts of other Acts to form the Employment Protection
(Consolidation) Act, 1978. This Act, together with others, has now
been amended by the Employment Act, 1980. Therefore, rather than
arrange this section of the chapter by individual Acts it has been
organised under topic headings.

6.5.1 Contracts of Employment

Subject to certain exceptions, within 13 weeks of starting work,
employees must be provided with a written statement giving details of
their job. This is not a written contract of employment and no signa-
tures are required. The main terms of the statement will have been
discussed already with the employee at his selection interview.

The written statement should include the information listed below; reference can be made to accessible documents such as an employees' handbook, if one is issued by the company.

(1) Rate of pay, method of calculation, and frequency of payment.
(2) Hours of work.
(3) Entitlement to sick pay, holiday pay, and pension.
(4) Period of notice.
(5) Job title.
(6) Disciplinary rules.
(7) The person with whom grievances can be discussed.

6.5.2 Itemised Pay Statements

With certain minor exceptions, all full-time workers should be provided with an itemised pay statement which should show the following details.

(1) Gross pay.
(2) Deductions and why these have been made.
(3) Net pay.
(4) If different parts of the net pay are paid in different ways, the amount and method of payment of each part should be shown.

Apart from the legal requirements, it is in the employer's interest to issue easily understood pay statements since these lessen the time spent answering queries on pay.

6.5.3 Payment of Wages

Manual workers, other than domestic servants, can demand to have their wages paid in cash. Statutory deductions such as income tax and National Insurance can be made by the employer; it is also possible, provided that the employee agrees, to make deductions for items such as food and accommodation.

Since 1960, wages can be paid by cheque or by credit transfer into a bank account, provided that a written request has been received from the employee. An employer, however, has the right to refuse an employee's request for a non-cash method of payment.

6.5.4 Equal Pay

The Equal Pay Act, 1970, which came into operation in 1975, requires that there is no discrimination in pay and conditions of work between men and women doing broadly similar work. If there is a dispute that cannot be resolved satisfactorily, the matter can be referred to an industrial tribunal. Arrears of pay for up to two years and damages may be awarded to the employee.

6.5.5 Sex Discrimination

The Sex Discrimination Act, 1975 is concerned with discrimination between the sexes in the field of staff recruitment, selection, training, and promotion. It is also concerned with discrimination between single and married persons.

Fair Discrimination

Discrimination is permissible only where there is a genuine occupational qualification, such as those indicated below.

(1) Specific physiological requirements such as those expected of artists' models.

(2) If there are considerations of decency; for instance, persons
working as lavatory attendants.

(3) Where people work and live closely together, such as the crew
of a small ship.

(4) Work in single-sex establishments such as prisons.

(5) Where statutory reasons prevent employment; for instance,
women should not work on night-shifts in factories.

Advertising Vacancies

It is unlawful to discriminate between the sexes in advertising
unless a genuine occupational qualification exists. The terms
'salesman', 'manager', and 'foreman' are, however, permissible provided that it is made clear that both sexes will be considered
equally for the post. Care must be taken over the use of 'he' or
'she' in the wording of advertisements.

Complaints and Enforcement

Specially prepared forms are available for those who think that they
have been discriminated against and guidance is available at employment offices on how to fill in the form. The forms are designed so
that questions can be put to the employer and complainants can be
helped to present their case to an industrial tribunal.

The tribunal can award compensation to an employee covering financial losses and expenses plus damages for hurt feelings.

6.5.6 Race Relations

The Race Relations Act, 1976, which replaced an earlier Act of 1968,
is complementary to the Sex Discrimination Act just described. It
has as its object the prevention of discrimination against racial
minorities even if the discrimination was unintentional.

Fair Discrimination

The genuine occupational qualifications allowed under this Act are
relatively few. It is, however, possible to select employees from
specific races on grounds of authenticity, such as for photographic
modelling or for serving customers in an ethnic restaurant.

Complaints and Enforcement

The arrangements to assist those who think that they have been discriminated against are broadly similar to those in the Sex Discrimination Act.

6.5.7 Redundancy

An employer can dismiss employees as redundant if there is no work
for them to do. Redundant employees are compensated by a lump sum
payment for their loss of employment and in consequence will be more
likely to give up their employment quietly rather than cause industrial unrest.

If employees are laid off or kept on short-time for long periods
they are, under certain circumstances, entitled to redundancy payments.

Coverage of the Act

With certain exceptions all workers are covered; the more important
exceptions are listed below.

(1) Those less than 18 years of age.

(2) Men over the age of 65 and women older than 60 years.

(3) Those who have worked for the employer for a period of less than two years.

(4) Some part-time employees.

(5) Crown servants, the police, and members of the armed forces.

(6) An employee who is either the husband or wife of the employer.

Alternative Employment

Should an employee refuse an offer of suitable alternative employ-ment, he may render himself ineligible for a redundancy payment. This offer must be made before his notice expires and may come from his existing or an associated employer or from the new owners of the business.

Payments

The size of a redundancy payment will depend on the length of contin-uous employment and the employee's rate of pay. The maximum payment is 30 weeks' pay, subject to a ceiling on the reckonable rate of pay.

Consultation

The employer must hold prior consultations with union representatives and notify the Department of Employment of proposed redundancies. Failure to start union consultations in advance of the redundancy may result in an industrial tribunal ordering an employer to pay the wages of redundant employees for a 'protected period' of up to 90 days. The lead-times for consultations are 90 days where more than 100 are to be made redundant and 60 days where the number exceeds 10. Failure to notify the Department of Employment may result in a fine or less assistance from the Redundancy Fund in meeting the redundancy payments. The Redundancy Fund is financed from National Insurance contributions.

Time off

A redundant employee with two or more years' service is entitled to reasonable time off to seek work or to arrange for training.

6.5.8 Unfair Dismissal

In general it is no longer possible to dismiss employees by giving them notice; the dismissal must in addition be fair.

Exclusions

The more important exceptions to protection by the unfair dismissal legislation are listed below.

(1) Employees with less than one year of service or two years if the number of employees has not exceeded 20 during the employee's service.

(2) Certain part-time employees.

(3) Employees who have reached normal retirement age. If there is no normal retirement age, this is taken as 65 for men and 60 for women.

(4) Employees who normally work abroad.

Fair Dismissal

If the employee claims unfair dismissal, it is the responsibility of the employer to show that the dismissal was fair. If any of the following conditions are met, a dismissal could be fair.

(1) Being incapable of doing the work after adequate training, proper supervision, and the provision of necessary facilities. The employee also should be warned and given an opportunity to improve his performance.

(2) Absence of appropriate technical or professional qualifications needed for the work.

(3) Being guilty of misconduct, examples of which include rudeness to customers, certain criminal offences, drunkenness, and refusal to obey reasonable instructions.

(4) Redundancy.

(5) Contravention of statutory requirements; for instance, a bus driver who loses his licence.

(6) Some other substantial reason. This could include unreasonable refusal to change working hours and false statements on an application form.

Constructive Dismissal

Should an employee leave his work voluntarily, even without notice, because of unreasonable treatment by his employer, this may be claimed to be constructive dismissal. If constructive dismissal can be proved, the employee will be covered by the unfair dismissal regulations.

Remedies for Unfair Dismissal

Should an industrial tribunal uphold a claim for unfair dismissal it will, provided that reinstatement is reasonable, attempt to have the employee fully reinstated in his old job, or in a comparable job elsewhere in the organisation. If the employer should unreasonably refuse to re-employ the person, an award can be made of up to 52 weeks' pay, subject to a maximum weekly amount.

In unfair dismissal cases, awards to employees normally consist of two parts. There is a basic award of damages of up to 30 weeks' pay and a compensatory award to cover losses incurred. Both awards are subject to limits that are reviewed annually.

6.5.9 Maternity

Expectant mothers are specially treated in employment legislation. They have four statutory rights: protection from dismissal on the grounds of the pregnancy, paid time off work for ante-natal care, maternity pay for six weeks, and a right to return at the end of a period of maternity leave.

To be eligible for maternity pay, a woman must satisfy three conditions. She must remain at work to within 11 weeks of the expected confinement. Secondly, she must notify her employer three weeks in advance that she will be absent from work as a result of the pregnancy. Finally, she must have worked for the employer for two years if a full-time worker, or five years if she works for between 8 and 16 hours per week.

If a woman is eligible for maternity pay, she can return to her old job, or another suitable one, provided that she properly notifies her employer of her intention to return to work. This right of return extends to 29 weeks from the date of her confinement. If an employee is denied the right to return, she can appeal to an industrial tribunal. The right to return does not apply to a woman who has worked for a firm having five or fewer employees.

6.5.10 Disabled Persons
The Disabled Persons (Employment) Act, 1958 is designed to assist disabled persons to secure work. Firms employing more than 20 persons must ensure that a least 3 per cent of their employees are registered disabled. This requirement need not be met if the firm has been issued with an exemption certificate. In selecting persons for certain types of vacancies, such as lift and car park attendants, preference must be given to applicants who are registered as disabled.

6.5.11 Employment of Ex-criminals
The Rehabilitation of Offenders Act, 1974 prevents certain criminal convictions being held against the offenders. To 'wipe the slate clean' they must complete a rehabilitation period without committing a serious offence during a period of from six months to ten years, depending on the severity of the sentence imposed. Rehabilitation does not apply to the medical and legal professions, to accountants, and to those who work with young people.

6.6 HEALTH AND SAFETY
Annually, hundreds of people are killed in British industry and hundreds of thousands of accidents of varying severity occur. Despite these figures, working in industry is still relatively safe and the risk of a fatal accident at home is almost as great as one at work. Millions of pounds are spent each year on the protection of industrial workers. In some sectors, such as the chemical and pharmaceutical industries, the marginal cost of saving additional lives is so great as to seem unjustifiable, when compared with the much lower cost of saving life by improved general preventive medicine.

6.6.1 Common Law Aspects of Safety
Apart from the statutory requirements of the Health and Safety at Work Act, there are common law requirements which, if neglected, can result in the payment of damages by the employer to the injured worker or his executors. The employer is required to provide a safe system of work; this can be analysed into a number of obligations. These are to provide each worker with the following.

(1) Competent fellow workers.
(2) A safe place in which to work.
(3) Working equipment that is safe to use.
(4) Adequate safety and protective equipment.
(5) Properly organised working arrangements.

6.6.2 Insurance Against Damages Awarded to Employees
The Employers' Liability (Compulsory Insurance) Act, 1969 makes insurance cover compulsory. It consequently ensures that any damages awarded to an employee can be paid, irrespective of the financial state of the employer.

6.6.3 Statutory Provisions - Health and Safety at Work
In 1974 the Health and Safety at Work Act became law. This enabling Act added to and embraced the existing legislation of the Factories Act, 1961, the Offices, Shops, and Railway Premises Act, 1963, and the Mines and Quarries Act, 1954. In addition, it brought under its jurisdiction employees such as postmen and teachers who had never previously been covered by any health and safety legislation; the

only major group now excluded is employees in private domestic service.

It is a criminal act for an employer to break the health and safety law; there is no limit to the size of the fine that can be imposed and imprisonment can be up to two years. The Act is not concerned with the award of damages, however. If the employee is dissatisfied with the award offered by his employer's insurance company, he can sue the employer in the courts.

The Health and Safety Commission

This body, which is appointed by the Secretary of State for Employment, has general responsibility for health and safety at work. It monitors the operation of the Act, promotes relevant research, and institutes inquiries. It also makes recommendations on regulations and codes of practice, the former having the force of law, the latter being advisory.

The Health and Safety Executive

This body is responsible to the Commission for the operation of the Act at local level. In general the work is performed by inspectors appointed by the Executive. If the premises are shops, offices, or catering establishments, the work is delegated to the local authorities. Inspectors are not new to the British industrial scene; the first factory inspectors were appointed in 1833. Within the Health and Safety Executive is the Employment Medical Advisory Service, which deals with all aspects of occupational medicine.

Enforcement

Inspectors of enforcing authorities have extensive powers of investigation. An 'improvement notice' can be served by an inspector when he considers that there has been a contravention of any statutory provision of the Act. This notice requires remedial action within a time limit. If an inspector considers that there is risk of serious injury he can issue a 'prohibition notice' which requires that there is no further use of the process until his recommendations have been carried out. Should an inspector have reasonable cause to believe that an article or substance is a cause of imminent danger, he can seize it and render it harmless.

Appeals against improvement or prohibition notices are heard by industrial tribunals.

Safety Representatives and Safety Committees

The involvement of employees in matters of safety is long established; since 1872 mineworkers have had the right to inspect mines. The present Act enables trade unions to appoint employee safety representatives who must be consulted in matters of safety. Safety representatives have the right to request that a safety committee be formed.

General Duties of Employers

Employers must, as far as is reasonably possible, ensure that they provide healthy and safe working conditions. They must also make adequate provision for the welfare of their employees. A written statement on health and safety policy must be produced and kept up-to-date. There must be a senior person who is made responsible for safety and health matters and this responsibility should be shared by managers and supervisors.

Duties of Manufacturers and Suppliers

Any person who designs, manufactures, imports, supplies, and installs anything for use at work has duties under the Act. He must ensure that it is safe when properly used and supply instructions concerning its correct use.

Duties of Employees

In matters of health and safety, employees are expected to take care of themselves and others. They must co-operate with their employers and must not misuse or interfere with anything provided for their health, safety, or welfare.

Factory Safety

Although by no means comprehensive, the following items are among the more important of those that might concern a Health and Safety Inspector when he visits a factory.

(1) The general register has been properly maintained and contains details of:

(a) Reportable accidents; that is, those that result in absences of longer than three days.

(b) Dangerous occurrences; for example, the bursting of a grinding wheel.

(c) Records of tests on hoists, steam boilers, and air receivers.

(d) A record of when the factory interior was painted.

(e) Names of authorised persons under the power-press and abrasive-wheel regulations.

(f) A record of testing fire alarms.

(g) A fire certificate.

(h) First-aid certificates.

(i) Certificates of approval of breathing apparatus.

(2) If work can be done sitting down, seats of suitable design are provided.

(3) There is no overcrowding. Each person should be provided with a minimum of 400 ft^3 (11.3 m^3) of space, with space ignored above a height of 14 ft (4.3 m).

(4) The level of lighting is adequate.

(5) The temperature in the factory is reasonable; 60°F (15.5°C) is required after the first hour where work is mainly sedentary.

(6) Adequate ventilation is available, with special provisions for the extraction of dangerous fumes and dust.

(7) Floors are of sound construction, well-maintained, and free from hazards.

(8) Sanitary arrangements meet the approved minimum scale and are kept clean.

(9) Appropriate provision is made for the washing and drying of hands.

(10) Accommodation is provided for employees' clothing with adequate drying facilities.

(11) Fencing of machines is substantial, properly maintained, and kept in position when parts are in motion.

(12) Special requirements applying to milling, grinding, and woodworking machines are observed.

(13) There are adequate means of escape in the event of fire. Audible fire alarms and fire-fighting equipment are provided.

(14) Adequate provision is made for first aid with trained first-aid staff.

(15) Meals are taken in separate rooms if poisonous substances are used in the manufacturing process.

(16) Ear protection is provided where sound levels are likely to damage employees' hearing.

(17) Eye protection is available and worn where appropriate.

(18) Chains, ropes and lifting tackle are of good construction and have been regularly examined and tested.

(19) The provisions applying to the employment of women and young persons under 18 years of age are being observed.

6.7 PATENTS

To encourage technical progress, the Crown can grant inventors exclusive rights to manufacture and sell their inventions. The period of protection is 20 years from the filing of the application for a patent with the Patents Office.

The Patents Act, 1977 is the latest development in British patent law. It gives greater protection for patents in the United Kingdom and aligns the law more closely with the patents law in other countries.

6.7.1 Patentable Inventions
To be patentable an invention must meet the following conditions.

Novelty

The invention should not be part of the 'state of the art' and therefore have not previously been available at any time or any place.

Inventive Step

The invention should be an inventive step, that is, it should not be obvious 'to a person skilled in the art'.

Industrial Application

It should be possible to use the invention in industry or agriculture. Substances used in a novel way in medicine and surgery can also be protected.

6.7.2 Items Not Covered
Scientific theory, mathematical method, computer programs, literary, dramatic, musical, and other artistic works are not considered to be inventions although some can be covered by copyright.

Novel medical and surgical techniques cannot be patented. Patents will not be granted for inventions that would encourage 'offensive, immoral or anti-social behaviour'.

6.7.3 Procedure for Obtaining a Patent
The procedure for obtaining a patent is long and expensive. It is advisable that the help of a patent agent should be sought at an early stage. Large companies are likely to employ their own patent specialists. The steps in obtaining a patent are outlined below.

Filing an Application

The first step will stake out a claim. The date of filing will establish a priority date for the application. An inventor with an earlier priority date will defeat another similar application with a later date. It is important that the inventor should not disclose

his invention, or offer it for sale, until he has filed his application for a patent.
   The application to the Patents Office should be accompanied by a
description of the invention, drawings, and a summary of its technical features. After the priority date has been granted it is possible to undertake further development work for a period of one year.
This work can be consolidated into the original invention, while
still using the original application date to establish priority.

Preliminary Application and Search
This stage occurs 12 months after the priority date and at the
request of the applicant. It enables the applicant to evaluate his
claim relatively cheaply and assists him to decide whether or not to
file his application abroad.

Publication
The first publication of the patent is in the *Patents Office Journal*
and occurs about 18 months after the priority date. Any infringement
of his patent between its first publication and the granting of the
patent is actionable. After publication, third parties may make
observations to the Comptroller of Patents on the patentability of
the invention.

Substantive Examination
The inventor will now have to decide whether he wishes to proceed
with the relatively expensive substantive examination. An examiner
employed in the Patents Office will then report on the suitability of
the invention for the granting of the patent. Should any non-compli-
ance be revealed, the inventor will be given an opportunity to amend
his application but will not be allowed to introduce new material.

Granting of a Patent
If a patent is granted, its life is 20 years from the filing date.
A patent is the personal property of its owner; it can be assigned to
others or mortgaged. Suspected infringements of patents can be
referred to the Comptroller who will decide whether or not an
infringement has occurred and determine damages. Civil proceedings
can be brought in the High Court which may issue an injunction
against the defendant and award damages to the plaintiff.

6.7.4 Ownership of Inventions
The employer, and not the employee, will own an invention if it has
resulted from the usual or assigned duties of an employee and could
be the normal output from his work. A significant factor in determining the ownership of a patent is whether the contract or job
description of the inventor requires innovatory skills. However,
should a patent that is owned by the employer be of outstanding
value, it is possible for the employee inventor to claim an award
from the Comptroller of Patents. This award may be of a lump sum of
money or of royalty payments.

6.8 MANUFACTURER'S LIABILITY

If a manufacturer produces a product that causes the death or injury
of a user, the manufacturer can be sued for damages. It is unlikely
that there will be a contract between user and manufacturer; hence
liability will be based on tort, where it must be proved that the

manufacturer has been negligent. For the case to succeed it must be shown that the manufacturer failed to take reasonable care, the injury was a direct result of the failure of his product, the designer and/or the manufacturer failed to take the precautions that the 'state of the art' required, and the designer should have taken into account the particular circumstances in which the accident happened. Under the present law the total cost of claims for damages against manufacturers is not high.

The present position could alter considerably if this country were to adopt laws on product liability similar to those in force in the United States. In 1976 an EEC draft directive was issued that accepted the principle of product liability but UK law has not yet been changed. If the law were changed it might no longer be necessary to prove that the injury was the result of the manufacturer's negligence but simply that the goods that caused the injury were unfit for the purpose for which they had been sold. As a result, the magnitude of the claims against manufacturers would increase, causing insurance premiums paid by the manufacturers to rise. This cost increase would be passed on in the price of the product, hence it would be the public who would ultimately compensate those users who suffer injury when using products.

Some of the awards made by the US courts are almost beyond belief. One quoted in 'Product Liability' by R.M. McRobb is of the 'lady who popped her poodle into her microwave oven for a few seconds to dry off after a bath. Before she could switch off the oven, the dog literally blew up in her face and she had a heart attack. But she got an award for damages against the maker of the oven'. Should Britain incorporate the principle of product liability into its law, it is to be hoped that a more sensible balance will be struck between the interests of the users and the manufacturers.

6.9 INDUSTRIAL POLLUTION

The extension of legislation on pollution is a reflection of greater public concern over protecting the environment. If a company infringes the law, it may face prosecution; if there is injury or damage, a civil action for compensation could result. Much of the earlier legislation on industrial pollution has been replaced by the Control of Pollution Act, 1974.

6.9.1 Pollution of Inland Waters

It is an offence for a company to discharge trade or sewage effluent into any inland watercourse or tidal estuary without the permission of the water authority. Permission of the water authority must also be obtained before similar effluent is discharged into the ground.

6.9.2 Waste Disposal

In England, the collection of general commercial and industrial waste is the responsibility of district councils and county councils are responsible for its disposal.

Particularly dangerous wastes are subject to special regulations, which require that records are kept and appropriate storage and transport facilities are provided. The services of specialist firms are available to deal with harmful wastes. Some larger companies install their own treatment plants. Dangerous wastes are often chemically treated, either to render them harmless or to convert them to a less

dangerous state. It can be a criminal offence to dispose of poisonous, noxious, or polluting waste in a way that is likely to create an
environmental hazard.

The United Kingdom Atomic Energy Authority is empowered to deal
with radioactive waste. It may also be concerned with the disposal
of other wastes that are as difficult and dangerous to deal with as
radioactive materials.

6.9.3 Air Pollution

The first legislation concerned with air pollution was the Alkali Act
of 1863; this was introduced to control the emission of hydrogen
chloride into the atmosphere. The emission of a wide variety of
noxious or offensive gases from 'scheduled works', which include
ironworks and steelworks, potteries, chemical works, power-stations,
and certain metal-smelters, is now controlled by the Health and
Safety at Work Act, 1974. The specified premises must be registered
annually and are subject to inspection. Should there be a contravention of the statutory provisions of the Act, improvement and prohibition orders may be issued.

The control of emissions from industrial premises, other than those
covered by the Health and Safety at Work Act, are dealt with by the
Clean Air Acts of 1956 and 1968 and the Control of Pollution Act,
1974. The enforcing authorities for these Acts are the district
councils.

The prolonged emission of dark smoke from industrial premises is,
with certain exceptions, an offence. Dark smoke is defined as darker
than shade 2 on the Ringelman chart. The emission of grit and dust
from furnaces in excess of prescribed limits is also an offence.
This problem is severe when pulverised coal is burnt and special
plant is needed to arrest dust and grit in the flue gases.

6.9.4 Noise

The Control of Pollution Act replaces most of the earlier legislation
on noise from industrial premises. If the local authority considers
any premises to be emitting an unreasonable volume of sound, it can
serve the occupier with a notice to abate the nuisance. Nuisance can
be broadly defined as noise at a level that causes substantial interference with health, comfort, and convenience. A common problem is
industrial noise affecting residential areas at night. Local authorities have the power to declare Noise Abatement Zones; in these
designated areas they attempt to stabilise and reduce noise levels.
If they wish, local authorities can control the level of noise coming
from construction sites; they did not have this power prior to the
Control of Pollution Act, 1974.

FURTHER READING

Day, N.J., Patents Act 1977, *Chartered Mechanical Engineer*, Vol. 25,
No. 6 (June 1978), pp. 77-80.
Fleeman, R.K. and Rhodes, R.J., *Employment Law: A Guide* (Fleeman
Cooper, Sutton Coldfield, 1981).
Henderson, J., *A Guide to the Employment Act 1980* (The Industrial
Society, London, 1980).
McRobb, R.M., Product Liability, *Chartered Mechanical Engineer*, Vol.
25, No. 5 (May 1978), pp. 75-77.

Ridley, J.R., **Engineering with Safety**, *Chartered Mechanical Engineer*, Vol. 28, No. 9 (Oct. 1981), pp. 52-54.

Vane, H.C., **Product Liability in the UK**, *Chartered Mechanical Engineer*, Vol. 30, No. 2 (Feb. 1983), pp. 58-61.

Walker, A., *Law of Industrial Pollution* (Godwin, Harlow, 1979).

Whincup, M., *Defective Goods* (Sweet and Maxwell, London, 1979).

7 Management of People

A large proportion of chartered engineers becomes managers and many
find the task of managing people more demanding than the technical
aspects of their work. This chapter starts with a discussion of
leadership and motivation; it concludes with an outline of the personnel function.

7.1 LEADERSHIP

There is little doubt that good leadership is essential in industry.
For senior posts, leadership of the highest order is needed; however,
at all levels down to junior supervision, leadership must be exercised.

7.1.1 Leadership Traits
Although great leaders of the past have varied widely in their
personal traits, it is possible to identify certain characteristics
that are likely to make a person a good leader. These include an
ability to organise and integrate the work of themselves and others,
decisiveness, self-assurance, persuasiveness, intelligence, initiative, intellectual honesty, integrity, and ambition.

7.1.2 Management Styles
A range of different management styles is possible. At one extreme
there is the autocratic leader who wields personal power, makes
decisions himself, and announces them as a *fait accompli* to his staff.
At the other extreme is the democratic leader who encourages his
staff to make as many decisions as possible and acts in a supporting
rôle. The classical school favours the more autocratic type of
leader whereas the behavioural school favours the democratic type of
leadership. A democratic form of leadership is likely to produce a
happier work-force, although it could result in reduced factory output.
   Some work situations are more suited to one style of management
than another; for instance, autocratic management is likely to be
appropriate to a stores but democratic management will probably be
more successful in a research department.

7.1.3 Behaviour of Leaders
Although some employees assume leadership rôles with considerably
more ease than others, leadership performance can be enhanced by
attention to the following points. Staff should be treated as intel-
ligent creative individuals; in fact, as people conforming to
McGregor's theory Y (see section 3.2.8). A leader should maintain
a high standard of personal conduct and not behave as 'one of the
lads'. He should keep his attitudes flexible and be ready to modify
his style of leadership to meet changing situations.

7.2 MOTIVATION

A manager must be able to get the best from his staff. This can be
achieved partly by the manager exercising his ability as a leader and
partly by designing work that will motivate employees.

7.2.1 Factors that Motivate

Work done by Maslow and Herzberg and already mentioned in chapter 3
is helpful when considering motivation.
   Maslow's hierarchy of needs were:

   1st Food and shelter
   2nd Security
   3rd Membership
   4th Esteem
   5th Self-fulfilment

   Once a lower level need has been reasonably satisfied, it ceases
to motivate and the next need in the hierarchy is aroused.
   Herzberg expanded on Maslow's list of needs and grouped them into
a two-level hierarchy. The lower level corresponded roughly to the
first, second and third levels of Maslow's hierarchy; Herzberg called
these 'hygiene factors' and indicated that they should be provided by
company policy and administration. When hygiene factors are absent,
employees feel 'exceptionally bad'. Herzberg used the word
'motivators' to refer to his higher-level needs and said that if
these were present, people felt 'exceptionally good'. Although money
is included as a hygiene factor, it also can be used to satisfy
higher-level needs. If sufficient money is available, a large house
and an impressive car can be acquired; these the owner usually assumes to be evidence of his achievement and of the esteem in which he
is held by his employers.
   It should be realised that individuals differ considerably in their
response to motivating factors. Some people lack ambition and have
little interest in their work; these are particularly difficult to
motivate.

7.2.2 Motivation and Job Design

Employees can be motivated by the way in which they are treated and
by the way their work is organised. Two aspects of work organisation
that can produce motivation are job enrichment and providing targets
for achievement.

Job Enrichment

This attempts to make work more interesting by changing the way in
which it is organised. Possible changes include greater freedom for
individuals and groups to organise their work routines, involving
employees in quality checks of their work, enhancing the job by
making it more skilled, and rotating tasks within a group of employees. In general, job enrichment should create a more contented
and co-operative work-force which produces work of higher quality.
However, there is likely to be a reduction in output when compared
with that obtainable when work is rigidly organised into simple
closely supervised tasks.
   Job enrichment will not motivate all employees; in particular it
will fail with those whose interests lie outside their work. Often,
employees prefer their work to be kept simple, provided that the

work-place arrangement enables them to chat with their neighbours while they work. An attitude that they often express is: 'Why convert a boring simple job into one that is both boring and complex?'.

Job enrichment appears to have the greatest benefit when applied to skilled and creative work and to work where a high standard of quality is necessary.

Providing Targets for Achievement

Staff can be motivated if they have objectives to attain. The over-all plan is to set company-wide objectives that are broken down into departmental and sectional objectives. Those who are expected to achieve these objectives should, where possible, be consulted before the targets are set. There should be frequent reviews of performance with longer feedback sessions at, say, six-monthly intervals.

It is often possible to associate a monetary incentive scheme with the achievement of objectives; for instance, the payment of bonus to sales staff who reach agreed targets. In repetitive production operations, employees often have targets set by work measurement. These imposed targets can be used as the basis for an incentive scheme based on output with daily monitoring of employee performance.

Motivation and the Manager

The individual manager can do much to motivate his staff by his conduct towards them. Firstly he should try to get to know them as individuals so that if he has to praise or rebuke them he can adjust his approach to the particular person. He should treat his staff with the respect and consideration that he would hope to receive if he were in their position. The manager should try to ensure that the necessary facilities are available for employees to do their jobs properly; if there is a failure to provide these he should be seen to be concerned. He should make it clear what level of performance he expects from his staff. If they fail to come up to his expectations he should discuss in private the reasons for their failure and en-courage them to do better.

7.3 THE PERSONNEL DEPARTMENT

While individual managers and supervisors lead and motivate their staff, the personnel department acts in a supportive rôle and advises on employment policies.

In particular, the personnel department is concerned with assistance in the following areas of work.

(1) Manpower planning
(2) Job analysis
(3) Recruitment, selection, and dismissal
(4) Education and training
(5) Working conditions, health, welfare, and safety
(6) Formal consultations with employees
(7) Job evaluation and staff appraisal
(8) Negotiations over wages and conditions of work
(9) Avoidance and the settlement of disputes

7.4 MANPOWER PLANNING

Labour is a resource that usually accounts for a substantial propor-tion of company expenditure. It is, therefore, important that human resources should be carefully planned and utilised. There are three stages in manpower planning: taking stock of existing resources, estimating future requirements, and producing a manpower plan.

7.4.1 Taking Stock of Existing Resources
The basis of this information is from personnel records which should
include the following information.

(1) Personal details - name, age, sex, marital status, address,
nationality, and next of kin.
(2) Work experience - previous employment and positions held in
previous and present employment.
(3) Education and training - formal qualifications held, courses
attended, and skills acquired.

It is helpful if existing employees are categorised. In larger
companies they can be grouped by site and then subdivided into occupational groupings. A grouping widely used is shown below.

(1) Managers
(2) Professionally qualified employees
(3) Supervisors
(4) Technicians
(5) Clerical employees
(6) Skilled manual employees
(7) Semi-skilled and unskilled employees

The age distribution of employees should be determined since this
is helpful as an indication of retirements. Other causes of labour-
loss will include death, disability, resignation, dismissal, and
redundancy.

Labour statistics can be usefully presented as trends plotted over
a period of, say, five years. Apart from presenting raw data, ratios
can be calculated to show, for instance, the ratio of direct to indirect employees, or the ratio of technicians to the total workforce.

Apart from predicted losses of employees, there are known manpower
inputs; these can come from apprenticeship schemes that supply
craftsmen and technicians.

7.4.2 Estimating Future Requirements
If there is a corporate plan (see section 2.4.6), this will no doubt
provide information on manpower needs over a one-year period and
possibly over the next five and ten years. If there is no corporate
plan, then the anticipated labour-load needed to meet the expected
levels of sales will have to be estimated. The direct labour requirements often can be calculated from standard times for manufacture and
assembly; these times should be adjusted for expected changes in
operator efficiency. If time standards are not available, labour
needs can be estimated from existing manning levels.

7.4.3 The Manpower Plan
This plan identifies the action needed to meet the manpower needs of
the corporate plan. The departmental manning requirements are compared with the expected manpower available so that shortages or surpluses can be identified. Hence, recruitment and training plans can
be prepared and, in the case of senior staff, detailed succession
plans drawn up. If labour surpluses are identified, arrangements
should be made to reduce the existing labour-forces.

The normal sources of new employees are those leaving the educational system, advertisements, 'head-hunting', Jobcentres and government training schemes. If the labour-force has to be reduced, first selectively cease recruiting and transfer surplus staff where possible, then encourage early retirements and finally declare redundancies.

7.5 JOB ANALYSIS

Job analysis examines individual jobs by looking at their component parts; from this information it is possible to prepare job descriptions and job specifications. Job descriptions are concise statements of what the job entails while job specifications indicate the qualities that are required from the people who do that particular job. Job analysis is therefore valuable not only for recruitment purposes but also for the training and assessment of staff.

7.5.1 Collection and Assembly of Information
A variety of methods is available to obtain information on job content. These include using a questionnaire completed by the person doing the job, interviewing the job holder and/or his supervisor, observation coupled possibly with the use of method study recording techniques (see section 9.3.3), and getting the person who does the job to complete a log of all work done during a given period.
The collected information about each job should be assembled in such a way that it facilitates the preparation of the job description and job specification. The following arrangement is an example of one that could be suitable.

(1) *Job Identification* This includes the job title, the work location(s), and the relationship of the job with those immediately above and below it in the organisation structure.
(2) *Work Content* The duties and tasks are stated here together with their purpose. Non-routine tasks should be identified and their frequency indicated.
(3) *Working Environment* Both the physical and the social environment should be recorded and unusual physical or mental demands noted.
(4) *Pay and Prospects* The basis on which the job is rewarded should be noted, together with any fringe benefits. The normal promotion route from the job should also be indicated.
(5) *Performance Criteria* The ways in which the performance of the job-holder is assessed and the standard of performance expected should be noted.
(6) *Special Job Characteristics* If there are factors about the job not already noted, these should be stated here.

7.5.2 The Job Description
This provides a concise word-picture of the job. It typically contains the following information.

(1) Job title
(2) Location of work
(3) Equipment and material used
(4) Supervision given and received
(5) Working conditions
(6) Special features of the job

Considerable skill is needed in writing the job description. The use of action-words to describe the job is recommended; examples of appropriate action-words are 'collects', 'analyses', 'manufactures', 'progresses', and 'estimates'. Further clarification of action-words may be needed; for instance, if 'collects data' is used, it could be desirable that there should be some indication whether this is done at the request of the supervisor or at the discretion of the job-holder.

Job descriptions are sometimes produced on standard pro forma sheets to achieve uniformity and to control length; a typical length is about 300 words. Once compiled, the job description should be acceptable to both the post-holder and to his supervisor; often it is signed by both parties. Updating is essential with the contents being reviewed and updated regularly. To avoid rigidity and to mini-mise argument, some job descriptions include a footnote indicating that post-holders will be expected to carry out other reasonable duties in addition to those listed.

7.5.3 Job Specification

This document provides a brief realistic statement of the human qualifications required to do the job. It is usually based on the opinion of the post-holder's supervisor and/or the job analyst. Typically the job specification includes information assembled under the following headings.

(1) Education
(2) Training
(3) Experience
(4) Judgement
(5) Initiative
(6) Skills
(7) Responsibilities
(8) Emotional requirements
(9) Special qualities

Some job specifications are compiled at two levels; the first spec-ifies the essential requirements, the second specifies the desirable requirements. For instance, in a given technician post the posses-sion of an appropriate Certificate of the Technician Education Council may be essential but the desirable qualification may be a Higher Certificate of the T.E.C.

7.6 RECRUITMENT OF EMPLOYEES

Recruitment is concerned with defining vacancies, determining the source of suitable candidates, and attracting their applications.

The definition of vacant posts is greatly facilitated by the avail-ability of job descriptions and job specifications.

7.6.1 Sources of Recruitment

Internal

This source should be considered first as it is quick, inexpensive, and the candidates will be well-known to the company. In addition, the morale of employees can be improved if they know that vacancies are filled, where possible, by internal promotion.

External

There are many sources of external recruitment; these include edu-

cational establishments, Careers Offices, Jobcentres, private
agencies, replies to advertisements, and the recommendation of company employees.

Private agencies can supply candidates for a wide range of vacancies. A field in which agencies are strong is in the provision of
clerical staff. Normally a charge is made on appointment and this
charge is refunded if the employee leaves within a given time.
Recruitment consultants specialise in the supply of senior management
staff. For a fee they will undertake the whole recruitment procedure
and present the employer with a short list of candidates from which
the final choice can be made.

7.6.2 Advertising Vacancies

Advertising vacancies is one of the most common methods of recruitment. The advertisement may consist simply of a board placed outside
the factory to indicate current vacancies and asking applicants to
contact the personnel department. Jobcentres also will advertise
vacancies if they are notified of them by employers. Newspapers and
professional journals are, however, the usual vehicles for job advertisements. Normally, local newspapers are used to advertise for
clerical and factory employees; the national 'heavies' are used for
management and senior technical staff. Professional journals are
often used to advertise technical vacancies.

It is suggested that the advertisements should contain the following information.

(1) Company name and type of business
(2) Job title and brief description of the work
(3) Location of the work
(4) Skills, qualifications, and experience required
(5) Remuneration and fringe benefits offered
(6) Action needed to make initial contact with the company

Non-display-type advertisements are only a little less successful
than more expensive display advertisements in attracting replies.
However, many companies prefer large display advertisements since
they feel that these are more appropriate to their public image.

Box numbers should not be used unless the company is launching a
new venture about which it wishes to preserve secrecy. Companies
that do not wish to disclose their identities often hire recruitment
consultants to advertise on their behalf. These consultants treat
all applications in confidence and will not forward applications to
any company nominated by the applicant thus avoiding the risk of an
applicant applying for his own job.

7.7 SELECTION

Selection normally involves compiling a short list based on information obtained from application forms, then interviewing the short-
listed candidates to decide whom should be appointed. Selection is
sometimes undertaken in two stages with selection tests and preliminary interviews being used to help compile the final short list.

7.7.1 The Application Form

This form is completed by all applicants; it conveys information in
a standardised manner which facilitates comparison between candidates.
Usually a company will use different forms depending on the broad
category of work being sought. After an appointment has been made,

the application form of the successful applicant is frequently used as part of the employee's personal record. Typical information collected by an application form includes

(1) Job that is sought
(2) Name of applicant, address, and telephone number
(3) Date of birth, marital status, and nationality
(4) Education and training record
(5) Medical history
(6) Employment history
(7) Signature of the applicant and date

7.7.2 The Short List
Using the information provided by the application form and any other information available, a short list of about five of the most promising applicants is prepared. If the selection process is in two stages, a primary short list of perhaps 10 or 15 persons is compiled.

7.7.3 Selection-testing
A wide variety of tests can be administered, although their use is by no means universal. The validity of certain types of selection test is in doubt; however, well-chosen tests can be helpful in sorting out the applicants. Three commonly applied tests will be described.

Intelligence Tests
Intelligence can be measured with reasonable accuracy in groups having similar backgrounds. It must, however, be appreciated that intelligence only is being measured; other equally important factors, such as effort or perseverance are not being measured.

Personality Tests
Personality is an important but difficult characteristic to measure. Apart from the selection interview itself, one of the most popular tests of personality is the leaderless group discussion. Here a group of candidates are asked to discuss a particular topic. Observers watch and listen but do not participate, other than to introduce the topic and to close the proceedings. They try to detect the candidates favourable traits such as patience, firmness, and courtesy, as well as unfavourable ones such as conceit, loquacity, and timidity.

Aptitude Tests
For some types of factory or office work it is desirable to determine whether applicants possess specific aptitudes. Standard tests are available to measure manual dexterity, spatial perception, and mechanical aptitude.

7.7.4 Selection Interviews
Some form of interview is inevitable; if well conducted it can be a useful, although not a certain, indicator of the best candidate. Interviews enable the personality of the candidate to be observed and the relevance of his experience to be examined. Conventional interviews are much more likely to be successful than stress interviews. In stress interviews the interviewer attempts to provoke and unsettle the candidate by his unusual behaviour.

The arrangements for interviewing can vary considerably. Single
interviews may be used; here only the candidate and the interviewer
are present. This type of interview can establish a good rapport
with the candidate but is liable to be affected by the personal
prejudices of the interviewer. An interview with a panel of about
three or four members should provide a wider range of expert questions and a concensus opinion of each candidate's suitability.
Rapport is, however, more difficult to establish with panel interviewers. A compromise between single and panel interviews is the
successive single interview. Here, typically three interviewers hold
single interviews with each of the candidates; they then meet to pool
their opinions.

Interviews should be carefully organised so that there is ample
time for all but the most talkative candidate. The interviewer
should have read the job description, job specification, and the
application forms. A note should be made of any points on which
further information is to be sought at the interview. When there are
panel or successive interviews the questions should be planned to
avoid repetitive or disorganised questioning.

After welcoming the candidate and indicating where he is to sit,
the interviewer should attempt to establish a rapport. This can
normally be achieved by adopting an easy friendly manner and listening with interest to what the candidate says.

It is a good idea to break the ice by talking about something non-
controversial such as the weather or possibly the journey. The
application form provides a basis for asking questions although a
plodding chronological review is to be avoided. An initial question
to a mature candidate about his recent work is likely to be much more
appropriate than delving into the recesses of his school record.
Probing questions should be asked about work experience and the
reasons for any gaps in the work history on the application form.
Other questions frequently asked concern what the applicant thinks
he can contribute to the company and what position he would wish to
occupy in five or ten years. Sufficient time must be left for the
candidate to be told something of the company, the job, and its
prospects. Also there should be adequate time for the candidate to
ask questions.

Unless applicants are to be told the result of the interview before
they disperse, they must be told how and when they will hear the
result.

Many interviewers make up their minds about the suitability of a
candidate too early in the interview, or even on the basis of the
application form. Another common fault is that the interviewer talks
too much. The applicant should be encouraged to do most of the talking; this can be done by creating a relaxed informal atmosphere and
by asking open-ended questions. The interviewer however must remain
in control, guiding the conversation into areas in which he is seeking information and regulating the pace of the proceedings.

Opinions differ on some aspects of interviewing. A number of companies favour closely structured interviews where the interviewer
asks a comprehensive standard list of questions. This approach,
however, can turn the interview into an interrogation and inhibits
a free flow of conversation. Another matter on which views vary is
note-taking. Some interviewers take notes only after the interview
whereas others make notes during the interview. If the latter course

is adopted the notes must be brief to avoid interrupting the contin-
uity and unsettling the candidate. Some companies provide specially
prepared forms for interviewers' notes; others encourage the use of
the application form for jotting down brief notes.

References are usually taken up. It is difficult to know how much
reliance can be placed on opinions expressed in character references
unless their author is known. Companies are unlikely to give unfavourable references in writing but they may be more frank over the
telephone.

Medical examinations are not usually required unless there is some
doubt about the applicant's health or the work makes unusual demands.

7.8 DISMISSAL AND RESIGNATION

Dismissal is the termination of the contract of employment by the
employer. Dismissal without notice is unusual and is justified only
where serious misconduct has occurred. If dismissal is with notice,
the appropriate period of notice depending upon length of service
must be given. There is a considerable body of legislation affecting
dismissal; this has been outlined in chapter 6.

Resignation is where the employee gives his employer notice that he
intends to leave. Some resignations are disguised dismissals where
the employee is given the option of resigning or being dismissed. If
an employee resigns but has no other work to go to, he may have
difficulty in obtaining unemployment benefit.

7.9 INDUCTION

The use of short induction courses for new employees was mentioned
in chapter 4. For most types of work a one-day course is usually
sufficient; ideally it should be held on the first day at work.

A senior manager should welcome the group and tell it something
about the company and its products. There should be a conducted tour
of the factory and offices, during which useful locations such as the
surgery, canteen, and wages office may be pointed out. Information
can be provided on safety, methods of payment, and any company sports
and social activities. Many companies provide booklets giving details of such things as sickness and holiday pay, superannuation,
safety, and timekeeping.

Finally, new employees should be taken to the sections in which
they are to work and introduced to their supervisors. The supervisor
should introduce them to their immediate colleagues, to the shop
steward, and to the safety representative. The supervisor should
tell new recruits about their job and explain how it contributes to
the work of the department and the company.

7.10 TRAINING

Some newly appointed employees will require training. Existing staff
also may have to be retrained, updated, or prepared for promotion.
Training may be in-house, at private or government training centres,
or in further or higher educational establishments.

7.10.1 Skill Training

This type of training usually occurs in-house. It is used for new
recruits and for existing staff who are to be transferred to new
work. Training can take from less than one hour for simple semi-
skilled work to several months for more highly skilled tasks.

Two methods of in-house skill training will be briefly outlined;

other methods such as skills analysis and the discovery method are
described in *Manpower Training and Development*, by Kenney, Donnolly
and Reid.

'Sitting with Nellie'

This traditional method involves watching an experienced operator and
copying her. Provided that Nellie is a good worker and the job can
be learned by watching and copying, this is a cheap and fairly effective method of training. Apart from its use in training production
operators this method is also used to train salesmen who accompany
experienced colleagues on their visits to customers.

Training within Industry

TWI is a long-established group of training programmes developed by
the Department of Employment. The best-known programme trains
employees on-the-job with the supervisor acting as the instructor.
After finding what the person already knows, the supervisor demon-
strates the job step-by-step, emphasising the important points relating to safety and quality. The person being trained then does the
job while the supervisor watches and corrects where necessary. This
procedure is repeated until the supervisor is satisfied with the
trainee's performance. The trainee is then left to get on with the
task and the supervisor returns as necessary to check that all is
well. TWI job instruction is a quick, cheap, and effective method of
training over a wide range of factory and office work.

7.10.2 Attitude Training

This form of training is used by some companies to encourage their
employees to become more co-operative and progressive in their attitudes. The training normally takes place off-the-job with the train-
ers using rôle-playing and case-study exercises.

   Middle and lower-grade managers from different parts of large companies are often brought together for several days on residential
courses designed to improve attitudes. Away from their working
environment, employees are likely to be more receptive and to participate more readily. A difficulty that is often found when attitude
training is conducted at residential establishments is that course
members become fired with enthusiasm only to be disillusioned when
they return to work.

7.10.3 Technical Training

Colleges of Further Education and Higher Education offer a wide range
of courses including those of the City and Guilds Institute, the
Technician Education Council, and the Business Education Council.
Many colleges, both in the public and private sectors, will arrange
courses on company premises provided that numbers are sufficient.
Some large companies have their own technical training schools.

7.10.4 Management Training

Staff selected for possible promotion can have their potential developed and assessed by suitable training programmes. Knowledge of
management techniques and management-related subjects can be broad-
ened by attending courses that prepare students for the examinations
of the Institution of Industrial Managers or for the Diploma in
Management Studies. For supervisors there are the courses of the
National Examination Board for Supervisory Studies (NEBSS).

Management is, however, best learnt by practising it and certain appointments of junior responsibility may be organised to provide experience where mistakes will not prove to be disastrous. Another method of providing training is by attachment as a personal assistant to a suitable senior manager.

Managers need to be able to communicate, persuade, and motivate; these interpersonal skills can be developed by special courses.

7.11 LABOUR TURNOVER

This statistic represents the rate at which employees leave the company.

$$\text{Annual labour turnover} = \frac{\text{Number of employees who left in a year}}{\text{Average number employed during year}} \times 100$$

A high labour turnover is expensive for, apart from the cost of procuring and possibly training a large number of replacements, it results in dislocation of production and a lowering of staff morale. High labour turnover can result from one or more of the following causes: poor selection methods, unsatisfactory working conditions, inadequate supervision, or uncompetitive wages. Misleading high labour turnover figures can be caused by new employees who stay for a few days only and they often relate to a comparatively small number of unpopular jobs where it is difficult to retain employees.

Labour turnover should therefore be read with labour stability to obtain a balanced picture

$$\text{Labour stability} = \frac{\text{Number of employees with more than one year of service}}{\text{Average number employed during year}} \times 100$$

7.12 WAGES AND SALARIES

Wages are paid weekly to manual employees and the payment is quoted in money per hour. Salaries are paid monthly to clerical, technical, and managerial employees and are quoted on an annual basis. Many consider this practice to be divisive and outdated; some companies treat all employees as salaried staff.

The basic rate of pay for hourly paid employees is usually augmented by additional payments. Examples of these additional payments include shift allowances, long service awards, and bonuses based on output.

7.12.1 Payment Related to Output

Many factories pay direct operators a wage that varies with their output. A well-designed incentive scheme can often raise output by about one-third as compared with that when operators are paid only a fixed hourly rate. It is suggested that incentive schemes for production operators should incorporate the following factors

(1) Wages should be proportional to effort.

(2) Effort should be measured using a sound system of work measurement.

(3) When a minimum wage is used it should be set sufficiently below average earnings to ensure that almost all of the employees can earn bonus.

(4) The reward should be paid as soon as possible after the effort.

(5) There should be some means of safeguarding quality standards.

(6) Employees should be compensated if they fail to earn bonus because of reasons beyond their control.

(7) The system should be easily understood.

There are a large number of payment-by-results wage-incentive schemes in use. A comprehensive treatment of them is provided in the ILO's publications *Payment by Results*. Three popular schemes are described below.

Piece-work with Guaranteed Minimum Wage

This method pays wages in direct proportion to output and incorporates many of the features of the model scheme described in the previous section. A graph of earnings against effort and the distribution of effort from a typical well-motivated group of employees is shown in figure 7.1a. It will be seen that with the minimum wage set at 75 British Standard (BS) effort, almost everyone will be earning a bonus. However, if the minimum wage is raised to 100 BS effort, only half of the employees would be earning bonus and in consequence the motivation of those not earning bonus would fall significantly. Effort-rating scales used to measure operator performance are explained in section 10.3.1. The steeply increasing direct labour cost per piece when performance falls below bonus level can be seen in figure 7.1b.

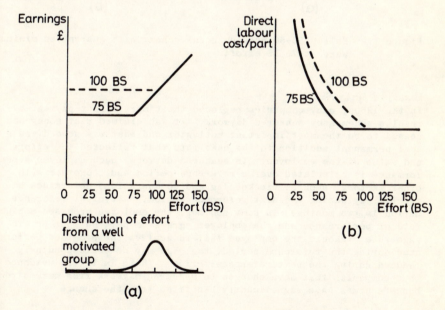

Figure 7.1   Direct piece-work with guaranteed minimum wage at 75 BS and 100 BS effort

Geared Piece-work Schemes

One of the best known of these is the Halsey 50-50 premium bonus scheme. Usually the target time is set at 75 BS effort and any time saved is shared equally between the employee and the company. It can be argued that this method of payment is unfair to employees since they are not fully rewarded for their effort. An advantage claimed for geared incentive schemes is that they smooth out earning varia- tions caused by inaccurate work measurement. The scheme is illus- trated in figure 7.2. Earnings plotted against effort are shown in figure 7.2a and the direct labour cost per piece is shown in figure 7.2b. Other incentive plans of this type pay employees different proportions of the time they save.

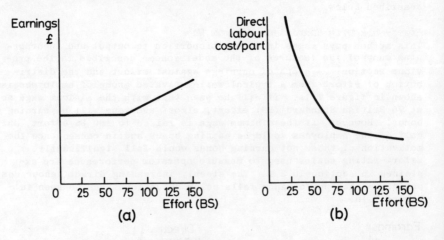

Figure 7.2   Halsey 50-50 premium bonus scheme with guaranteed minimum wage at 75 BS effort

Measured Daywork
In the 1960s, many companies replaced their more direct payment-by- results schemes by measured daywork. It was claimed that money had ceased to be the most important motivator and what was needed was a semi-permanent addition to the base rate that reflected the effort and value of the employee. In measured daywork, each employee's per- formance is calculated over a reference period and, together with other factors, such as timekeeping and co-operation, determines his rate of pay during the next reference period. A typical reference period is two months. In some schemes, the employee promises a high rate of performance and the employer agrees to pay the appropriate wage rate. Should the employee fail to achieve the promised perform- ance during the reference period, the wage may be correspondingly reduced during the next reference period. There is little evidence that companies that have changed to measured daywork from more direct bonus schemes have significantly benefited from the change.

7.12.2 Group Bonus Schemes
Incentive schemes based on output are suitable for groups as well as individuals. Group schemes reduce administrative costs, create co- operation and flexibility within the group, and thereby reduce the work of the supervisor. Care must be taken to choose reasonably small coherent groups; if the groups are too large, each person feels less personally involved and group output is likely to be lower.

Service employees, such as production control and maintenance sta
are often paid a lieu bonus which reflects the performance of the
production groups that they serve.

7.13   JOB EVALUATION

Job evaluation provides a rational system of determining the pay of
various jobs and hence creates a logical and defensible pay structure.
It would, however, be incorrect to claim that job evaluation is a
scientific method of determining pay since all methods rely on human
judgement. Despite its objectivity, trade unions are often reluctant
to accept the findings of job evaluation. Before any evaluation is
attempted, employees should have the principles of job evaluation
carefully explained to them.

An evaluation committee should be formed; a typical committee would
consist of the personnel officer and the head of management services
together with departmental managers and union representatives as
appropriate.

Job descriptions and job specifications should be available for use
by the committee.

7.13.1   Method of Job Evaluation

A wide choice of methods is available; most methods are described in
the *Guide to Job Evaluation* by Incomes Data Services and Livy's *Job
Evaluation: A Critical Review* (see Further Reading at end of this
chapter). In this section two of the traditional and one of the more
recent methods will be outlined.

Classification

This semi-analytical method constructs a hierarchy of grades into
which jobs are placed. Between five and eight grades are used each
with its appropriate salary range. The grades are carefully defined
in terms of skill, training, and responsibility. If a wide range of
jobs are to be classified, they should be grouped into a number of
compatible occupational groups each with their own grade definitions.

Initially, a number of well-known representative jobs are classified; these are referred to as 'key jobs'. Job specifications are
used by the evaluation committee to help it to decide into which
grade each of the key jobs should be placed. After the committee has
satisfied itself that the evaluation of the key jobs is satisfactory,
they slot the rest of the jobs into their appropriate classifications.

This method of job evaluation is quickly applied and easily understood. It is best used for stable jobs as it is relatively inflexible to technical and organisational changes.

Points-rating

Points-rating is the most widely used technique of job evaluation; it
is often called the 'weighted-points method'. Jobs are analysed into
a number of factors and a marking scheme is devised; each factor is
awarded a maximum number of points in the marking scheme depending on
its relative importance. Normally five factors are used; skill,
physical effort, mental effort, responsibility, and working conditions. If necessary these can be increased in special situations by
adding other factors such as creativity or danger. Since factors
have different importance in different occupational groupings, for
instance, physical effort is more important in manual than in clerical work, it is usual to have different marking schemes for different
occupational groups.

The committee starts by selecting a number of key jobs in one of
the occupational groups. It then awards points to the factors in
each key job according to the marking scheme. To facilitate the
award of points, factors are normally subdivided into a number of
grades, each with an appropriate points award. The points score for
each key job is then totalled and the points awarded plotted against
present job earnings. This graph, referred to as a scatter diagram,
is shown in figure 7.3. A line of best fit is drawn through the
points on the scatter diagram; this line acts as a points-to-money

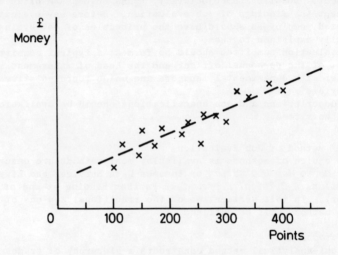

Figure 7.3 Scatter diagram showing points-to-money conversion line

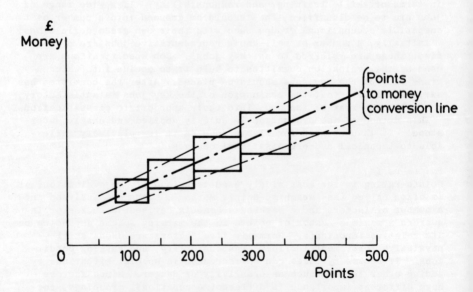

Figure 7.4 Stepped points-to-money conversion graph

conversion line as well as indicating which jobs are relatively over-
paid or under-paid. After consideration of national and local wage
rates and a review of company wages policy, it may be decided to draw
a revised points-to-money conversion line. To avoid too many wage
rates, the points range is often subdivided into a number of job
grades as shown in figure 7.4.

After the key jobs have been evaluated, the committee evaluates the
other jobs and the points it awards are used to determine appropriate
wage rates.

Despite its relative complexity, the points method of rating provides a logical and stable system of job evaluation which is favoured
by some trade unions.

Guide Chart Profile

There are a number of recently introduced systems of job evaluation;
one of the most popular is the Hay-MSL Guide Chart Profiling method.
It was originally intended for management and higher-level technical
staff but its use has now been widened.

Jobs are examined under three factors: know-how, problem-solving,
and accountability. These factors are subdivided and points awarded.
Ranking is then used to check the points awarded and to determine the
final grading of the jobs. An attraction of this type of job evaluation is that the consultants, Hay-MSL, provide an up-dated link
between gradings and salaries being paid nationally.

7.14 STAFF APPRAISAL

Although supervisors should be continually appraising individuals on
their staff, many companies consider it desirable that there should
be a formal annual assessment. These appraisals enable wage increases to be justified for individual employees and for staff to be
selected for promotion. If employees are informed of their assessments it also can act as a useful motivator. The term 'staff appraisal' tends to be used to describe the assessment of senior staff;
merit-rating is normally used when lower-level staff are assessed.
Some trade unions object to the application of merit-rating to their
members.

7.14.1 Merit-rating

A merit-rating form is used to record and help in the assessment of
personal characteristics such as job-knowledge, accuracy, work-speed,
co-operation and initiative. Each characteristic is typically graded
into five factors varying from exceptionally good to inadequate.
Assessments are usually made by the immediate supervisor on an annual
basis. Most companies interview employees and tell them the results
of their merit-rating; some inform them only if their rating has been
unfavourable. Merit-rating will not operate satisfactorily unless
supervisors are prepared to allocate extreme ratings, where these are
justified, and to avoid personal prejudices when making assessments.

7.14.2 Appraisal Interview

Annual appraisal interviews between senior staff and their managers
are usually less formal than the merit-rating procedure just described. The interviewer discusses individually with each staff member his performance and encourages him to make self-criticisms. At
the interview, the manager should have the moral courage to criticise
frankly individual shortcomings and to indicate how performance can
be improved.

7.15 INDUSTRIAL RELATIONS

7.15.1 Trade Unions

These organisations represent about half of the working population of
the United Kingdom. The four main categories of trade union are outlined below.

Craft Unions

This type of union is descended from the medieval trade guilds and
consists of members who possess specific trade skills. Some craft
unions, such as the National Graphical Association, still retain
their craft traditions; others, such as the Amalgamated Union of
Engineering Workers, have ceased to represent exclusively skilled
workers. Craft unions have a horizontal structure representing their
members irrespective of the industry in which they are employed.

Industrial Unions

Here the unions aim to represent all those employed in a particular
industry whatever work they do. Two examples are the National Union
of Mineworkers and the National Union of Railwaymen; both of these,
however, have failed in their aim of securing a complete representation of the industry they claim to serve.

Occupational Unions

These draw their support from those in a particular occupation. An
example is the National Union of Teachers which will accept school
teachers as members but not dinner ladies or caretakers.

General Unions

There are no occupational, industrial, or craft restrictions in
general unions. The largest union in the United Kingdom, the Transport and General Workers Union, is an example of a general union.

   The Trades Union Congress (TUC) is a voluntary body to which most
trade unions are affiliated. The Congress meets annually; here
resolutions from constituent unions are debated and policy on industrial and political matters is determined. The General Council of
the TUC acts as a voice of the trade union movement; it may be consulted by the government and will mediate in disputes between unions.

   The trades union movement is a democratic one. Individuals are
members of a branch that is associated with the union at national
level, either directly or through a district level. The union executive committee and full-time officials are usually elected at annual
union meetings. Owing to the apathy of the average union member,
branch officials, and conference delegates are often elected by an
unrepresentative minority of political activists.

   Full-time union officials help to solve local problems and represent members in national collective bargaining. Local bargaining
usually takes place between management and shop stewards with full-
time officials being called upon if their help is needed. Local
agreements made by shop stewards can considerably enhance the pay and
conditions determined by national agreements. Shop stewards are company employees who are elected to represent groups of employees; they
usually spend part of their working day on union matters. The chairman of a factory shop stewards' committee is often called the works
convenor; he often works full-time on union duties.

7.15.2 Employers' Organisations

At national level the unions negotiate with the appropriate employers' association. National agreements are concerned only with
basic conditions of employment and minimum rates of pay for certain
broad categories of employees. Not only do national agreements protect employees by providing a floor below which wages and conditions
of employment cannot fall, they also protect employers from competitors in the United Kingdom who might otherwise be encouraged to reduce production costs by paying low wages.

Employers' associations are industry-based voluntary bodies that
represent the interests of their members. An example is the Engineering Employers' Association. These associations are formed to
promote the commercial interests of their constituent companies and
to negotiate with trade unions. The Confederation of British Industries (CBI) is a voluntary confederation of employers' associations
and has as its principal purpose the promotion of the prosperity of
British industry. In many ways the CBI is the counterpart of the
TUC.

7.15.3 Disputes

Although strikes represent avoidable loss of output and sometimes
cause much public consternation and inconvenience, the days lost by
strikes in the United Kingdom are considerably fewer than those lost
as a result of industrial accidents. Most strikes are unofficial;
that is, they are called by shop stewards without official union
backing. Official strikes are called only after the agreed procedure
for settling disputes has been exhausted.

Official Procedure

Procedures for settling disputes differ from industry to industry but
the steps described below are typical. The next step is tried only
if the earlier one fails to produce a settlement.

(1) The shop steward raises the grievance with middle management.
(2) A meeting is held between the district officer of the union and
senior management of the company.
(3) There is a discussion between union officials and the regional
committee of the employers' association.
(4) At national level there is a meeting between the union and the
employers' association.

In some industries there is also an agreement to use an arbitrator
if the fourth stage does not provide a solution.

Local Procedure

A number of sanctions can be used by shop stewards to try to settle
a local dispute in their favour. These are indicated below in order
of escalation.

(1) Withdrawal of co-operation with management.
(2) Work to rule.
(3) Restricting output by going slow and refusing to work overtime.
(4) Withdrawal of labour, often for short periods of time.

Although in a dispute employers can apply sanctions such as refusing to allow overtime to be worked or locking workers out of the

factory, their usual sanction is to refuse the demands of the shop stewards.

FURTHER READING

Bell, D.W., *Industrial Participation* (Pitman, London, 1979).
Betts, P.W., *Supervisory Studies* (Macdonald and Evans, Plymouth, 1980).
Bramham, J., *Practical Manpower Planning* (Institute of Personnel Management, London, 1978).
Fraser, J.M., *Employment Interviewing* (Macdonald and Evans, Plymouth, 1978).
Graham, H.T., *Human Resource Management* (Macdonald and Evans, Plymouth, 1980).
ILO, *Payment by Results* (International Labour Office, Geneva).
Incomes Data Services, *Guide to Job Evaluation* (Incomes Data Services, London, 1979).
Kenney, J., Donnolly, E., and Reid, M.A., *Manpower Training and Development* (Institute of Personnel Management, London, 1980).
Livy, B., *Job Evaluation: A Critical Review* (George Allen and Unwin, London, 1975).
Taylor, L.K., *Not for Bread Alone: An Appreciation of Job Enrichment* (Business Books, London, 1980).

8 Financial Management

The growth and survival of a company depends on its profitability and solvency. The engineer works with the accountant to achieve these ends; it is therefore important that the engineer understands the procedures and vocabulary of the accountant. Financial and secretarial responsibilities are normally combined under the financial director whose duties can include the following.

(1) Advising the board on financial and legal matters.
(2) Planning and controlling expenditure in accordance with company policy.
(3) Administering budgetary control.
(4) Designing and operating management information systems.
(5) Collecting and paying money, keeping books of accounts, and preparing the annual accounts.
(6) Maintaining statutory records concerning wages, salaries, and National Insurance and fulfilling legal responsibilities under the Companies Acts.
(7) Arranging for the issue and transfer of shares, payment of dividends, and notification of shareholders' meetings.
(8) Assisting in capital investment appraisal.

Accounting has tended to divide into two main divisions: financial and management accounting.

Management accounting is concerned with the preparation and presentation of accounting information to assist management in the formulation of policies and in the planning and control of the company. In doing this, management accountants work in close co-operation with factory management.

Financial accounting is largely concerned with the cash inflows and outflows of the company. The more important of these cash flows together with the outside groups involved are shown in figure 8.1.

8.1 ANNUAL REPORTS AND ACCOUNTS

The annual report of a company consists of a balance sheet, a profit and loss account, and various items of supporting information. It provides a record of the stewardship of the board of directors over the past year. The Companies Acts require that in the annual report certain minimum information be provided for shareholders. The Acts also require that the annual accounts are audited by independent auditors.

8.1.1 Balance Sheet

This is a statement of what the company owes and owns at the end of its financial year. The major components of a balance sheet are

123

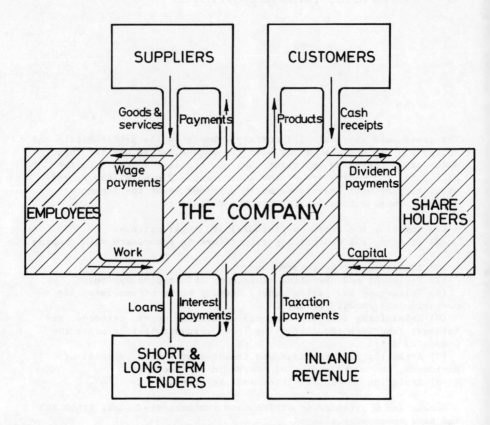

Figure 8.1   Major inflows and outflows of manufacturing company

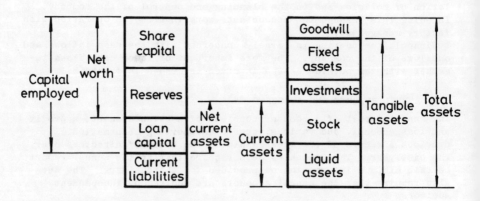

Figure 8.2   Balance sheet terminology

shown in figure 8.2; the right-hand column indicates the total assets
and the left-hand column indicates the shareholders' claim and the
company's liabilities. The terms used in figure 8.2 are expanded
below.

(1) Share capital - issued ordinary and preference shares.

(2) Reserves - retained profit.

(3) Loan capital - money lent to the company for a fixed term.

(4) Current liabilities - money owed by the company to trade
creditors, the bank, shareholders, and the Inland Revenue.

(5) Goodwill - an intangible asset representing customer goodwill
to the company.

(6) Fixed assets - land, buildings, machinery, and vehicles owned
by the company.

(7) Investments - money that has been invested outside the company.

(8) Stock - the value of materials, work in process, and unsold
products.

(9) Liquid assets - debtors (those who owe money to the company)
and cash.

A typical balance sheet of a large manufacturing company, as it
might appear in its annual report, is shown below.

BALANCE SHEET - As at 31st July 198-

ASSETS EMPLOYED   £m

Current assets (A):	Cash		25
	Debtors		255
	Stock		250
			530
Current liabilities (B):	Creditors		200
	Bank overdraft		25
	Taxation due		25
	Dividends due		20
			270
Net Current assets (A - B)			260
Fixed assets			400
Investments			130
			790

FINANCED AS FOLLOWS

Issued share capital	250
Reserves	420
Total shareholders funds	670
Loan capital	120
	790

8.1.2 Profit and Loss Account

In this account, the income from the goods supplied during the year
is matched with expenditure on resources required to achieve that
income. Profit is the difference between income and expenditure.
After tax has been paid on this profit some of the balance is dis-

tributed as dividends to shareholders; the residue is retained in the business to increase the reserves.

Profit is not simply cash received less cash paid out in a given period. Income is realised when goods are supplied to customers and not several months later when payment is normally received. Similarly, not all of the relevant resource expenditure will have been paid for by the end of the period. Secondly, expenditure must be matched with income. This means that, except on long-term contracts, only expenditure incurred on goods sent to customers in the period under consideration should be charged. Hence, inventory associated with orders not yet delivered is excluded from expenditure and appears as an asset on the balance sheet.

The profit and loss account that corresponds to the balance sheet shown in section 8.1.1 appears below.

PROFIT AND LOSS ACCOUNT - For year ended 31st July 198-

		£m
Sales		1100
Cost of goods sold		950
Trading profit		150
Interest from investments	30	
Loan and overdraft interest (deduct)	20	10
Net profit		160
Taxation		60
Net profit after tax		100
Dividends paid or recommended		30
Profit retained		70

The profit and loss account and balance sheet items are considerably amplified in the 'Notes on the Accounts', which often occupy several pages in the annual accounts.

8.1.3 Accuracy of Annual Accounts
Annual accounts cannot be fully accurate even though they have been correctly prepared.

A major cause of balance sheet error is in the statement of asset value. Assets subject to wear and tear are normally shown at a proportion of their original cost that reflects their unexpired life; this reduction in value with time is called depreciation. The most popular way of calculating annual depreciation is the straight-line method where annual depreciation = (installed cost - terminal value)/ estimated life. Both the estimated life and the terminal value of an asset can be wrongly estimated. An even more important cause of inaccuracy in times of inflation is the distortion of asset values caused by the reducing value of the currency unit in which the assets are stated. This applies particularly to the longer-term assets, such as plant, land, and buildings; here periodic revaluations are necessary if current values are to be quoted. An asset difficult to quantify is goodwill. This is an intangible asset that can arise from the favourable opinion in which the company is held by its customers; in addition it can be caused when other companies have been bought at a cost in excess of their tangible assets. Goodwill does not always appear in balance sheets. Finally, it should be appreciated that the balance sheet is a statement of worth on one day

of the year; the position could have changed considerably by the time
the report is published.

The profit and loss account can also be misleading. Anticipated
losses on work due for completion in future financial years are subtracted from profit in the current year, whereas future profits are
not similarly anticipated. Profit can also be directly affected by
any errors in the valuation of stock or work in progress.

In times of high inflation, annual accounts can be grossly misleading if prepared using conventional accounting principles. Accounting
methods that recognise inflation have been developed. Larger companies have, for several years, published balance sheets and profit
and loss accounts based on current costs, in addition to conventional
accounts. Current-cost accounting is a matter of considerable controversy within the accountancy profession. Inflation accounting is
discussed in section 8.1.6.

8.1.4 Value Added

Many companies now include a statement of value added in their annual
report. Value added indicates the wealth created by the company's
trading effort during the year. It is obtained by subtracting from
sales revenue the cost of goods and services purchased and adding
investment income. The statement analyses the use of wealth created;
this often indicates the large proportion of value added paid to
employees as wages and salaries. A value-added statement that could
accompany the examples of balance sheet and profit and loss account
given in sections 8.1.1 and 8.1.2 is shown below.

VALUE-ADDED STATEMENT - For year ended 31st July 198-

		£m
Sales		1100
Less materials and services purchased		500
Value added		600
Add investment income		30
Total wealth created		630

Used as follows:

To employees pay, national insurance, and superannuation		420
To government as taxation		60
To providers of loans		20
To shareholders as dividends		30
		530
Retained in business		
Depreciation	30	
Retained profit	70	
		100
		630

8.1.5 Analysis of Performance

The annual accounts can be analysed to provide a guide to company
performance. Most measurements are expressed as non-dimensional
ratios that tend to eliminate the effect of changing currency values

although it should be appreciated that the effects of inflation on profits are seen earlier than on asset values. Ratios can be plotted over a number of years to establish trends; they can also be compared with those of similar companies. Some measures of performance and financial status are indicated below.

Profitability

This is measured as (profit/capital employed) × 100. It shows the return on capital employed and indicates how effectively assets are being used. The terms in the ratios should be unambiguously defined; a suitable definition of profit is pre-tax profit plus loan interest and capital employed is issued share capital plus reserves and loan capital. A favourable value in engineering manufacture would be of the order of 25 per cent.

The main profitability ratio can be subdivided into two subsidiary ratios

$$\frac{\text{profit}}{\text{capital employed}} = \frac{\text{profit}}{\text{annual sales}} \times \frac{\text{annual sales}}{\text{capital employed}}$$

Liquidity

If a company, irrespective of its profitability, runs out of money and cannot raise new money, it risks failure and subsequent liquidation. A primary measurement of liquidity is the current ratio where current ratio = current assets/current liabilities.

A commonly stated value of this ratio is two although many successful companies operate at a lower ratio. Much depends on the nature of the business and working-capital control. It will be appreciated that stock is included in current assets. However, it is often difficult to convert stock into cash rapidly. Hence, another ratio, which takes account only of quickly realisable assets, is used; this is the 'acid test' ratio: acid test ratio = (current assets - stock)/current liabilities. This ratio should not be less than one.

Stock Turnover

This is the number of times that the balance sheet figure of stock can be divided into the cost of annual sales. The size of this ratio will depend on the type of production and, to some extent, the level of activity. A high figure is desirable; about four is typical for a large-scale batch-production factory in the engineering industry.

Debtor Ratio

The debtor turnover is found by dividing the figure for debtors into the annual sales. A high figure is desirable and one of around five is typical. The average debt-collection period can be found by dividing the figure for debtors by the average daily sales. In addition to trend comparisons, this ratio can be compared with the average time that the company is taking to pay its creditors.

Wages/Sales

If the total wages bill is stated in the annual report, this ratio can be calculated. There will be considerable variation with type of business and degree of automation. A high ratio indicates that the company is vulnerable if wage rises cannot be fully recovered in increased selling prices.

Interest Cover
This ratio can be obtained by dividing the profit before loan interest and tax have been deducted by the annual loan interest. A low
ratio means that the ordinary dividend and the share price are at
serious risk if profits fall and, more seriously, that the company
may not be able to meet the loan interest it has contracted to pay.

Dividend Cover
If the total amount available for distribution to the ordinary shareholder is divided by the amount actually distributed, the dividend
cover is found. A high ratio is likely to indicate a large plough-
back of money into the business and that the ordinary dividend rate
could be increased.

The following ratios cannot be obtained from the annual accounts
but are widely used as measures of company performance.

Net Price/Earnings (PE) Ratio
This is found by dividing the net earnings per share (profit per
share after deduction of tax and preference dividends) into the current ordinary share price. A high figure normally indicates that the
company's prospects are highly rated by investors.

Dividend Yield
The dividend paid is divided by the current ordinary share price. A
low ratio normally indicates that the company is held in high regard
by investors.

8.1.6   Inflation-accounting
Engineers normally use units that remain constant in value. However,
the primary unit of the accountant, the pound sterling, varies in
value with the passage of time. As already stated in section 8.1.3,
the varying value of the unit of currency can result in misleading
annual accounts. In times of inflation, conventional (historic cost)
accounts tend to overestimate profit and undervalue assets; these
distortions can become dangerously misleading when the rate of inflation is high. In the inflationary 1970s, considerable efforts
were made by the accountancy profession to find a suitable method of
allowing for inflation in company accounts. Although the annual
accounts of companies are still prepared on an historic-cost basis,
larger companies now provide additional information on the effect of
inflation on their results. This information is in the form of a
current-cost profit and loss account and balance sheet.

Current-cost Profit and Loss Account
In this account, three adjustments are made to the historic-cost
trading profit to arrive at the current-cost trading profit. The
first of these is an increased depreciation charge; this reflects the
greater value of fixed assets when valued at their current replacement cost. Secondly, there is a cost-of-sales adjustment which takes
into account the additional cost of replacing stock at current prices
rather than at its historic cost. The third adjustment, called the
'monetary working capital adjustment', is made to take account of the
impact of price changes on the amount needed to maintain the monetary
working capital of the business. Monetary working capital consists
principally of the difference between money owed by debtors and the

money that is owed to creditors. When the total of the current-cost adjustments is subtracted from the historic-cost trading profit, the current-cost trading profit is found.

Further work is needed to find the current-cost profit before taxation. Firstly, a gearing adjustment is made to eliminate that part of the current-cost adjustments that have been funded not by the ordinary shareholders of the company but, for instance, by the holders of loan stock. The other deduction is no different from that appearing in the historic-cost profit and loss account; this is the interest payable to loan-stock holders, less the dividends obtained from investments.

The effect of current-cost accountancy on the profit of a leading British company can be seen from the 1982 accounts of Racal Electronics.

		(£ thousands)
Trading profit on historic-cost basis		114,393
Current-cost adjustment		
Depreciation	14,444	
Cost of sales	13,290	
Monetary working capital	5,106	32,840
Current-cost trading profit		81,553
Gearing adjustment	(11,537)	
Interest payable, less investment income	11,777	240
Current-cost profit before taxation		81,313

In 1982, Racal Electronics' current-cost profit per share was 18.22 p, against 26.16 p when calculated on an historic-cost basis. In fact, some companies, particularly low-geared companies with substantial fixed assets, find that their dividend is not covered by their current-cost profit. This produces in company reports remarks such as: 'Dividends paid and proposed in respect of 1981 are not covered by profit for the year determined in accordance with SSAP 16 procedures, but retentions of current cost from previous years are sufficient to meet the deficit'. SSAPs are Statements of Standard Accounting Practice issued by the accountancy profession; these help members to narrow the scope for opinion on topics where a greater degree of information is considered desirable. The SSAP dealing with current-cost accounting is No. 16.

Current-cost Balance Sheet
The amount stated here for fixed assets is the current cost of their replacement, based on valuation or cost, updated by appropriate indices. This is in contrast to their treatment in historic-cost accounts where fixed assets are normally stated at their original cost, less accumulated depreciation. Also appearing in the current-cost balance sheet is the Current Cost Reserve; this shows the net effect of all current-cost adjustments on balance-sheet items.

8.1.7 Auditing
From 1900, Companies Acts have required that the annual accounts of public companies be checked by independent auditors. The object of auditing accounts is to discourage fraud and to try to ensure that shareholders are presented with accounts, based on stated accounting

conventions, that truly represent the state of the company.

The auditor's report forms an essential part of a company's annual
report and accounts. Almost always their report is favourable but if
the auditors are unhappy they qualify their report. Unless the
qualification is on a minor technical point, qualification is a
serious matter which could adversely affect the company's reputation
and share price.

Appointment of Auditors

In the first instance auditors are appointed by the directors. However, at each annual general meeting of the company, the shareholders
are asked to sanction the re-appointment of the auditors and to authorise the directors to fix the auditor's remuneration. An auditor
cannot be removed from his appointment without the shareholders'
agreement. Auditors must meet the qualifications laid down by the
Companies Acts; in practice they are firms of accountants.

Independence of Auditors

To do their job impartially, auditors should be completely independent of the company whose accounts they are examining. To this end
it is desirable that the partners and staff of the auditors should
have no financial interest in the companies they audit. Secondly,
to avoid commercial pressures, the audit fees from any single company
should not constitute more than perhaps five per cent of the auditor's
gross income. Also, the auditors should not provide other fee-earning
services, such as taxation advice, to audit clients.

8.2 EXTERNAL SOURCES OF FINANCE

In addition to the initial cost of setting up a business, most companies require additional capital from time to time. This is used to
finance expansion or to help the firm over periods of difficulty. In
this section, capital structure and external sources of finance are
discussed.

8.2.1 Share Capital

The two major types of share capital are ordinary and preference
shares with the former being much more important.

Ordinary Shares

This type of share does not receive a fixed rate of dividend; the
dividend policy is recommended by the board of directors. Although
ordinary shares are issued at a nominal value, their actual value
depends on the current profitability of the firm and investors'
opinion of its prospects. Large companies have their shares bought
and sold on the Stock Exchange, with the price quoted daily in the
financial columns of newspapers.

Ordinary shares entitle their holders to a share in the residue of
company profits after all other distributions have been made. Ordinary shareholders normally have voting rights at company meetings in
proportion to their holding of shares. Owing to inflation, the size
of ordinary dividend distributions has in general tended to increase,
often keeping pace with the rate of inflation.

When an established company wishes to raise new money by the issue
of additional ordinary shares, it can do so by a rights issue. Here
existing holders are offered the right to buy a specified number of

shares at a price below the current market price. If a holder does
not wish to purchase his allocation of shares he can sell the rights
to purchase to another person.

Preference Shares

Preference shares normally carry a fixed annual dividend; they have
preference over ordinary shares in the allocation of profit for dividends and to compensation in the event of liquidation. Variations of
standard preference shares include those that are convertible into
ordinary shares, those on which dividend is cumulative in the event of
a missed or reduced dividend, and those that will be redeemed at the
end of a fixed period.

Like ordinary shares, the preference shares of larger companies can
be bought and sold on the Stock Exchange. Unlike most ordinary
shareholders, owners of preference shares are not normally entitled
to vote at company meetings.

Owing to their fixed rate of dividend, preference shares have for
a long period been unpopular with investors. To the company they are
less attractive than loans since dividends have to be paid out of
taxed profits whereas loan interest is paid before tax.

8.2.2 Loan Capital

Loans usually carry a fixed rate of interest and are repaid at the
end of a given period of time. Loan capital is normally a smaller
proportion of the total capital of a company than share capital. The
relationship of loan and preference capital to capital represented by
the ordinary shares is referred to as gearing. A company with a high
proportion of fixed interest capital is referred to as highly geared.
Such a company in times of rising profit finds that it can pay dis-
proportionally increased dividends to its ordinary shareholders but,
if trading conditions deteriorate, there may be nothing left for them
after loan interest and preference share dividends have been paid.
Loan interest must be paid irrespective of profitability.

Debentures

This type of loan is usually secured against specific company assets,
such as land. Debentures can have a fixed life, or may be irredeem-
able during the life of the company.

Loan Stock

Large well-regarded companies have this form of financing available
to them. The invitation to subscribe can be to the general public,
financial institutions, and existing shareholders. Convertible loan
stock is a popular variant in times of inflation since holders can
convert their stock into a pre-arranged number of ordinary shares at
some future date. In the event of liquidation, owners of loan stock
are treated as trade creditors and hence have less security than
debenture holders.

Term Loans

These are made by financial institutions. They are a convenient way
of raising money for smaller companies that do not have access to the
stock market. Sometimes the lender may require a directors' guarantee. This guarantee enables the lender to sue the directors to the
full extent of their private assets in the event of a failure to meet
the terms of the loan.

Overdrafts

Bank overdrafts are a popular and relatively cheap method of borrowing money. If firms use overdrafts to finance their stocks and
debtors, this form of borrowing can be considered part of the company's capital. Overdrafts are negotiated for an amount of money up
to an agreed limit. Interest is charged only on the current size of
the overdraft and at a rate of a few percentage points above the minimum lending rate; interest paid is allowable as a charge against
tax. In the event of liquidation, banks normally have no legal
claims against specific assets owned by the company.

8.2.3 Financing by Hire Purchase or Leasing

When a specific item of plant or equipment is required, it is sometimes financed either by hire purchase or by leasing. In the case of
leasing, the property remains in the ownership of the lessor whereas
with hire purchase the property passes to the user on completion of
instalment payments. The comparison of the cost of hire purchase and
leasing is complex and is influenced by the availability of investment grants and by tax considerations.

8.3 INTERNALLY GENERATED FUNDS

These are a valuable source of company finance and provide established companies with a substantial source of funds. The two most
important methods of generating funds internally will be described.

8.3.1 Retained Earnings

Instead of distributing as much earnings as possible to ordinary
shareholders, most companies usually retain a considerable proportion
of their profits for reinvestment in the business. The longer-term
interests of the ordinary shareholder will not be harmed by this
retention provided that reinvested funds do not dilute the earning
ability of the company.

8.3.2 Depreciation

Depreciation is another way of putting money back into the company.
It represents the reduction in value of tangible assets, such as
machine tools, due to use or obsolescence. This is recognised by a
periodic conversion of some of the asset's value into an expense
during the period of its useful life. If depreciation charges were
not made, or if these charges were inadequate, short-term profits
would increase at the expense of the longer-term profit-earning
ability of the company. A popular conventional method of depreciation
is the straight-line method which was described in section 8.1.3.
Conventional methods of depreciation will not generate adequate funds
to replace worn-out plant in times of inflation and methods of depreciation that take account of expected replacement cost have been
devised.

8.4 ANALYSIS OF PRODUCT COST

The selling price of a manufactured product can be analysed as shown
in figure 8.3.

8.4.1 Prime Costs

These are the costs that can be seen to be incorporated in the product as it is made and consist of direct material, direct labour, and
direct expenses. Direct material is the material used in the finish-

ed product. Direct labour is the labour involved in manufacturing
the product. Direct expenses are the other costs directly chargeable
to the product; for example, any subcontracting cost.

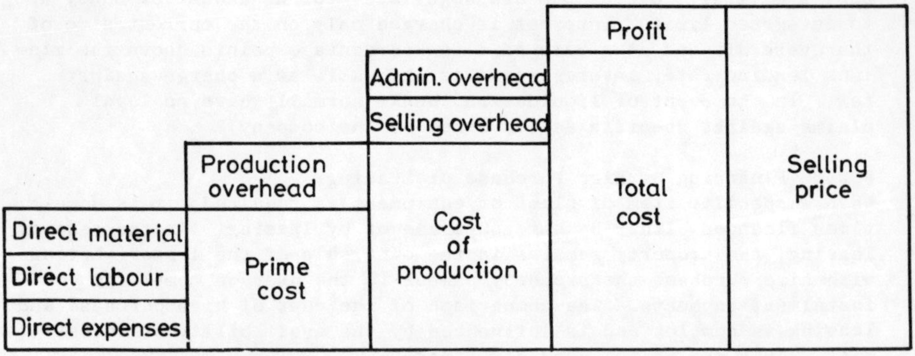

| Direct material Direct labour Direct expenses | Prime cost | Production overhead | Cost of production | Selling overhead Admin. overhead | Total cost | Profit | Selling price |

Figure 8.3   Analysis of product cost

8.4.2   Production Overheads
The cost of production, other than the prime cost, is termed production overheads. These include factory maintenance, power, lighting
heating, works management, supervision, storekeeping, internal transport, and depreciation charges on the production facilities. These
costs can be divided into two components: those that are fixed and
those that vary with output.

8.4.3   Selling Overheads and Administration Overheads
Selling overheads consist of the cost of distributing and marketing,
including advertising and warehousing. Administration overheads are
the costs incurred by the company that are not directly associated
with either selling or production.   These overheads also can be
divided into fixed and variable components.

8.4.4   Profit
The size of the addition to the total cost to provide a profit is a
policy decision in which management is advised by the sales department. Although this addition will often be as much as the market
will stand, sometimes a low profit or even a loss is planned on a
particular product in order to penetrate a new market or to obtain a
greater share of an existing market.

8.5   ALLOTMENT OF OVERHEAD COSTS
Prime costs can be directly attributed to specific products but overheads are of a more general nature. Despite their general nature,
these costs must be recovered in the prices charged for the products.
Unless a correct allotment of overheads is made to products, it is
impossible to know true product costs and in consequence incorrect
selling prices may be charged.

8.5.1   Recovery of Production Overheads
The sequence of steps used to allot production overheads to products
is outlined below. Here, both fixed and variable overhead costs are
recovered.

(1) Allocate to departments those overheads that are wholly identified with them, for example, supervision wages.

(2) Apportion other overheads to departments on an equitable basis.
An example of this primary distribution is lighting costs, which can
be apportioned on the basis of the floor area of each department.

(3) Re-allot the totals of the service departments' costs to production departments so that, as the result of this secondary distribution, all overhead costs have been loaded on the production departments.

(4) Charge the various production departments' overheads from stage
3 to jobs or products on a basis that best reflects the way in which
the production overhead is used by the products. Four bases of
charging production overheads are

    (a) Direct labour cost
    (b) Direct labour hours
    (c) Machine hours
    (d) Unit of product

If the direct labour cost basis is chosen, then the direct labour
cost for each department will have to be estimated for a period of,
say, one year. This is then divided into the estimated annual departmental overhead cost from stage 3 above; hence a departmental-
percentage direct labour cost overhead recovery-rate can be found.

The cost centres for the recovery of production overhead need not
be departments; in fact, when the type of equipment used in departments varies widely, the machine-hour recovery basis is more appropriate. Here a cost centre is a group of similar machines and a
production overhead recovery-rate is an appropriate hourly charge for
each group of machines on which the products are manufactured.

8.5.2 Recovery of Selling Overheads and Administration Overheads
Selling overheads can usually be allotted to specific products,
enabling them to be recovered as appropriate percentage additions to
the cost of production. Often there is no attempt to allocate administration overheads to individual products and they are recovered as
a standard percentage of the cost of production.

8.6 ABSORPTION COSTING AND MARGINAL COSTING
The method of costing in which all of the overheads are recovered as
an addition to the prime cost is termed 'absorption costing'. There
will be over-recovery or under-recovery of overheads if actual output
differs from that estimated and profit projections will be affected.

Overhead costs can be divided into fixed and variable components.
A system of costing called 'marginal costing' ('direct costing' in
the United States) does not attempt to allocate fixed costs to individual products but allows them to be recovered as contributions from
all the products. The contribution made by a product is the difference between its selling price and its total variable cost; a negative contribution represents an under-recovery of variable cost.
Profit is made when the total of all the contributions exceeds the
fixed cost. The following simple example illustrates the difference
between absorption and marginal costing.

Marginal Costing

£ thousands	Product			
	A	B	C	Total
Sales	100	50	100	250
Variable costs	65	45	70	180
Contribution	35	5	30	70
Fixed costs	-	-	-	50
Profit	-	-	-	20
Contribution/sales	35%	10%	30%	28%

Absorption Costing

	Product			
	A	B	C	Total
Sales	100	50	100	250
Variable cost	65	45	70	180
Fixed costs	25	5	20	50
Total cost	90	50	90	230
Profit	10	nil	10	20

It will be seen that although product B does not make a profit, it still makes a contribution. If sales can be increased, product A is to be preferred for, although it makes the same unit profit as C, its contribution is greater. Marginal costing is useful when spare factory capacity is available since it enables a price to be quoted for new orders that at least covers the variable cost. The question of over-recovery or under-recovery of overheads does not arise with marginal costing since fixed costs are not allocated to products. Marginal costing, although valuable when examining specific projects, is not widely used for regular management accounting reports.

8.7 BUDGETARY CONTROL

Budgets express in quantitative terms the operational targets of a company for a given period - usually one year. Budgets are based on the longer-term operations of a company which can be derived from its corporate plan (see section 2.4.6).

Regular comparisons are made between actual performance and budget. The variances thrown up by these comparisons are analysed and should promote appropriate corrective action and greater efficiency.

8.7.1 Preparation of Budget
The accountant is responsible for co-ordinating the work of budget preparation; he frequently does this through a budget committee made up of senior managers. The relationship between the major budgets is shown in figure 8.4 and will be outlined below.

Sales Budgets and Production Budgets
The sales budget is usually the first to be established. The production budget is derived from the sales budget with adjustments being made for proposed changes in the stock budget.

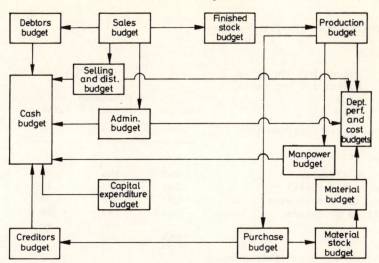

Figure 8.4   **Typical relationships of budgets in a manufacturing company**

Cost Budgets
The cost budgets are derived from the sales and production budgets. These budgets represent the costs that will have to be incurred to satisfy the proposed levels of sales and production. The total cost budget is analysed into budgets for labour cost, material cost, and overhead cost.

Departmental Operating Budgets
These budgets are a breakdown of the main budget to operational level. It is important that the managers concerned are involved in the preparation of their own budgets thereby making use of their valuable local knowledge and helping to secure commitment to budget targets.

Capital Expenditure Budget
Provision must be made for the replacement and the possible expansion of existing plant and buildings; this is done in the capital expenditure budget. Methods of assessing the benefits of expenditure on plant are dealt with in section 8.11.

Cash Budget
This budget is compiled after the completion of other budgets and predicts the cash available to the company during the budget period. Not only the size but also the timing of the expected cash flows have to be taken into account. The management of working capital is discussed in section 8.10.1.

8.7.2 The Control System
Budgets are of little use unless actual performance is compared with budget projections and the necessary corrective action taken by management. Annual budgets are broken down into either monthly or thirteen four-weekly control periods; against these targets, actual costs or performances are compared. A simple budget report is shown overleaf.

BUDGET REPORT for January 198-

Output

    Budget 100 000 units
    Actual 100 000 units

	(£ thousands)		
	Budget	Actual	Variance
Sales value of production (X)	1200	1190	10	A
Production costs			
Direct labour	200	198	2	F
Direct material	300	310	10	A
Variable overheads	100	95	5	F
Fixed overheads	150	140	10	F
	750	743	7	F
Selling and distribution costs			
Variable	40	38	2	F
Fixed	100	98	2	F
	140	136	4	F
Administration costs			
Variable	20	19	1	F
Fixed	80	81	1	A
	100	100	-	-
Total costs (Y)	990	979	11	F
Profit (X - Y)	210	211	1	F

A = Adverse variance
F = Favourable variance

Analysis of Variances
It will be seen from the table that some of the variances are favourable (F) and others are adverse (A). It is desirable that the
reasons for these variances should be explored. This can be done by
analysing the variances by department and by cause. For instance,
there could be an excessive usage of material in the press shop because a storekeeper is booking out phosphor bronze strip to production but dishonestly selling it to a scrap-metal merchant.

8.7.3 Flexible Budgetary Control
Budgets are prepared at a given level of output. In the example in
section 8.7.2 the actual and budgeted volume of output coincided.
However, when output is at a level different from that in the budget,
many costs will change and it is realistic to adjust the budget.
Certain costs will vary directly with output whereas others tend to
remain static. Hence the variable items such as direct material,
direct labour, and variable overheads will be adjusted to take
account of differences between actual and budgeted output.

8.8 STANDARD COSTING

The main activities of the business have their targets determined by
budgets. This target-setting process is extended in standard costing;
here target costs for products and their individual manufacturing
operations are determined in advance. These targets or standards are
normally set so that they are achievable under efficient operating
conditions. Deviations from target are recorded as variances which
can be analysed and so enable management to concentrate its attention
on areas of inefficiency as each arises. In this respect, standard
costing has a great advantage over historical costing which does not
provide targets and usually provides information too late for corrective action. Standard costing is widely used for repetition work; it
can also be used in jobbing work provided that standards can be
synthesised from elements of previous work or from standard data.

8.8.1 Standards

Accurate standards are usually determined for direct labour and
direct material costs with production overhead costs being controlled
by departmental overhead budgets and recovered by one of the methods
mentioned in section 8.5.1. Direct material costs are based on the
standard amount of material needed to manufacture the product, including material unavoidably lost in the shaping process. The direct
material cost is obtained by multiplying the standard quantity of
material by the standard material cost. Direct labour cost is found
from the standard or allowed time for the job, adjusted for the
target level of operator performance. Much of the information used
to compile direct labour and material costs is common to that used on
orders issued by the production control department; an example of a
component cost data sheet is shown in figure 8.5.

COMPONENT COST DATA SHEET					Part no. 265842 Description Bearing bolt				
Material 20 mm 0.4% carbon bright drawn M.S.	**Cost Code** 427		**Mat. Weight** 0.56 kg		**Std Price/kg** 180 p		**Std Mat. Cost** 101 p		
Operation	Dept Code	Lab Grade	Std mins	Std Labour Rate (p/h)	Std Labour Cost (p)	Std O/H Rate (%)	Std O/H Cost (p)	Labour + O/H (p)	
10 turn	61	M3	8.3	200	27.7	220	60.9	88.6	
20 turn	61	M3	3.4	200	11.3	220	24.9	36.2	
30 mill	52	F3	1.2	200	4.0	250	10.0	14.0	
40 drill	51	F3	1.5	200	5.0	270	13.5	18.5	
50 grind	66	M4	2.4	250	10.0	240	24.0	34.0	
60 bench	74	M3	3.6	200	12.0	180	21.6	33.6	
70 –									
80 –									
90 –									
100 –									
					70.0		154.9	224.9	
Summary	Material 101.0 Labour 70.0 Overhead 154.9 _____ 325.9								

Figure 8.5   Component cost data sheet

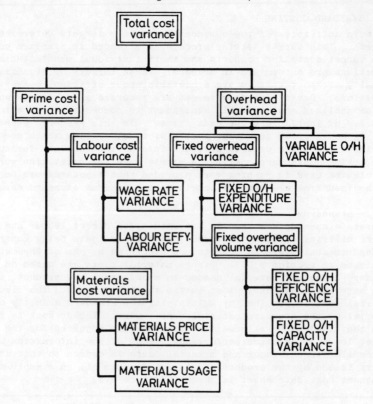

Figure 8.6 A breakdown of total cost variance

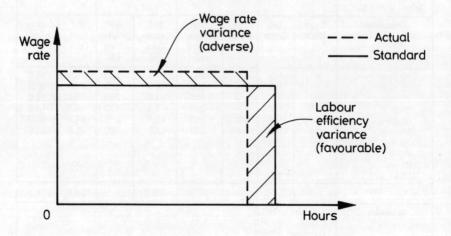

Figure 8.7 Direct labour cost variance

8.8.2 Variances

These are the differences between actual achievement and the targets set by the standards. A range of variances that is commonly used is shown in figure 8.6. These variances are obtained from the following expressions

(1) Total cost variance = Actual cost - standard cost

(2) Labour cost variance = Actual labour cost - standard labour cost

    (a) Wage rate variance = Actual hours × (actual rate - standard rate)

    (b) Labour efficiency variance = Standard wage rate × (actual hours - standard hours)

(3) Materials cost variance = Actual material cost - standard material cost

    (a) Materials price variance = Actual usage × (actual price - standard price)

    (b) Materials usage variance = Standard price × (actual usage - standard usage)

(4) Variable overhead variance = Actual variable overhead - standard variable overhead

(5) Fixed overhead variance = Actual fixed overhead - standard fixed overhead

    (a) Fixed overhead expenditure variance = Actual fixed overhead - (budgeted hours × fixed overhead recovery rate)

    (b) Fixed overhead volume variance = (Budgeted hours - standard hours produced) × fixed overhead recovery rate

        (i) Fixed overhead efficiency variance = (Actual hours worked - standard hours produced) × fixed overhead recovery rate

        (ii) Fixed overhead capacity variance = (Actual hours worked - budgeted hours) × fixed overhead recovery rate

The way in which wage rate and labour efficiency variances account for the difference between the actual and standard direct wages costs is shown in figure 8.7. Direct material variances also act in a similar manner.

The operation of the various fixed overhead variances can be seen from figure 8.8. The amount of fixed costs absorbed at different levels of capacity is shown by line OX. At point X the factory will be working at its budgeted capacity and the whole of the fixed overheads will be recovered. The actual fixed overhead expenditure is represented by point Y. These variances provide managers with useful information.

The efficiency variance indicates the under-recovery or over-recovery of fixed overheads caused by operators working at a pace different from that expected. Capacity variance shows the effect on overhead recovery of working at capacity other than that which had been budgeted.

It is usual to provide senior management with a monthly statement that indicates sales income, standard cost of production and cost variances. Actual manufacturing profit can be obtained by adjusting the standard manufacturing profit by the total variance. An example of a standard cost statement is shown overleaf.

STANDARD COST STATEMENT for January 198–

Output		Hours	
budgeted	10,000	budgeted	5,000
actual	10,200	actual	4,950

		£
Sales value of output 10,200 × £5 (A)		51,000
Standard cost of output (B)		
Direct labour	10,900	
Direct material	20,100	
Variable overhead	1,700	
Fixed overhead	3,300	36,000
Standard manufacturing profit (C) = (A – B)		15,000

Variances	Favourable	Adverse
Direct labour		
Rate	–	200
Efficiency	100	–
Direct material		
Price	–	120
Usage	–	100
Variable overhead	–	80
Fixed overhead		
Expenditure	–	–
Efficiency	20	–
Capacity	80	–
	200	500
Variance (adverse) (D)		300
Actual manufacturing profit (C ± D)		14,700

Production supervisors should be provided with regular statements of how their sections of the factory have performed against target. This statement should concentrate on items of expenditure within the control of each supervisor. A typical departmental operating-sheet for a foreman is shown in figure 8.9. In some companies, the bonus earned by a production foreman will depend on his performance against budgeted targets.

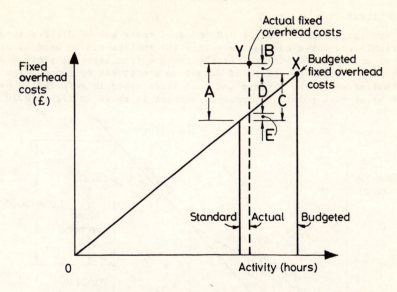

Figure 8.8 Fixed overhead cost variance. (A) Fixed overhead
            variance; (B) fixed overhead expenditure variance;
            (C) fixed overhead volume variance; (D) fixed overhead
            capacity variance; (E) fixed overhead efficiency
            variance

DEPARTMENTAL OPERATING SHEET						
Foreman Mr W. Jones			Dept Assembly		Week 18	
Output Budget 6548 Achieved 97% Actual 6361				Analysis of Labour Losses (£)		
	Budget (adjstd)	Actual Result	Variance		Responsibility	
					Direct	Indirect
Direct Operators				Excess costs		15
No directs	63½	60	− 3½	Rectification	100	5
Clock hours	2729	2714	− 15	Waiting	31	19
Wages paid (£)	2747	2597	− 150	Awards	−	−
Piecework perf. (%)	100	106	+ 6	Overtime	78	−
Labour losses (£)	250	280	+ 30	Overrated labour	−	−
Indirect Operators				Daywork	32	−
				Training	−	−
No indirects	3	3	−			
Indirect hours	121	132	+ 11	*Rating*		
Wages paid (£)	581	602	+ 21			
Indirect Materials				Output	Labour Utilisn	Quality	Over All
Small tools (£)	58	50	− 8			
Consble material (£)	105	120	+ 15	97	93	70	87
Tool Maintenance				*Notes*		
Material and Labour (£)	72	61	− 11			
Quality						
Scrap (£)	60	95	+ 35			
Rectification (£)	52	115	+ 57			

Figure 8.9 Departmental operating sheet

8.9 BREAK-EVEN ANALYSIS

As was indicated in section 8.6, product costs may be divided into
variable and fixed components. This information can be used to plot
total cost against output. If the revenue from sales is plotted on
the same graph, the effect of output on profit can be found. The
output at which the revenue and cost lines cross is referred to as
the break-even point. A break-even chart is shown in figure 8.10.

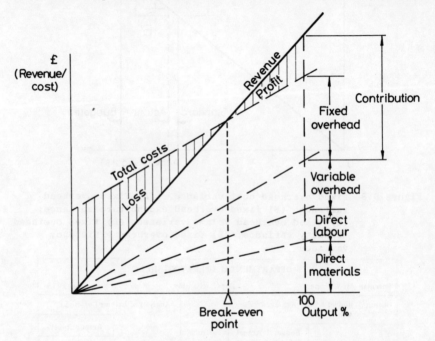

Figure 8.10 Break-even chart

   This type of analysis is useful when attempting to find the effect
on profit of relatively small changes in output. Caution should be
exercised when larger changes are involved; for instance, if output
drops considerably, fixed overheads are cut in an effort to maintain
profitability.
   Break-even charts clearly illustrate the greater vulnerability of
companies that have high fixed cost when output has to be reduced.
The two break-even charts in figure 8.11 show companies with equal
profitability and sales. Company B is highly capital-intensive
(large fixed costs but low variable costs) whereas company A has a
lower investment in plant and is more labour-intensive (small fixed
costs but high variable costs). It will be seen that to put company
B into a loss-making position, a much smaller drop in sales is required than for company A, that is $M_B < M_A$, where M is referred to as
the 'safety margin'.

8.10 MANAGEMENT OF CAPITAL

The capital employed by a company can be divided into two parts:
fixed and working capital. Fixed capital consists of land, buildings,

and equipment; its management is a long-term task since it changes
relatively slowly in most established companies.

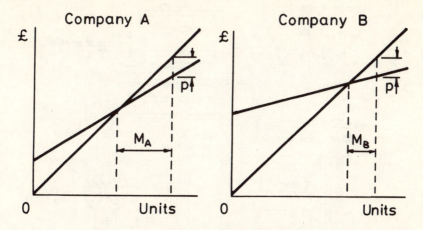

Figure 8.11   Break-even charts for companies operating at different
               levels of capital intensity

8.10.1 Working Capital

Working capital, that is, net current assets, is the amount by which
current assets exceed current liabilities. The management of working
capital requires constant attention if a satisfactory operating
balance is to be maintained.

   The circulation of cash in a manufacturing company can be seen in
figure 8.12.   The normal flow of working capital follows the manufacturing cycle, starting with the use of cash to purchase materials
needed in production.   Further expenditure is required on wages and
to meet the general running expenses.   These cash outflows are converted into work in progress and then into finished goods.   Goods are
distributed at a price that is increased by an addition for profit.
When customers pay their accounts, sales are converted into cash to
complete the cycle of working capital.   The only part of this cycle
not under company control is that between the despatch of goods to
customers and the receipt of payment.   Money can be syphoned off
working capital for the purchase of capital equipment.   Working capital is also reduced periodically when dividends, loan interest, and
tax is paid.   If required, there can be step increases in working
capital from time to time when money is raised on a temporary or
permanent basis.

   The amount of working capital should be maintained at its lowest
prudent level by minimising current assets and maximising current
liabilities.   Current assets can be kept low by carefully controlling
stock levels and by obtaining early settlement of customers' accounts.
The magnitude of current liabilities can be increased by taking
maximum advantage of the credit offered by suppliers of goods and
services.

   If surpluses of cash arise they should be skimmed off and put into
short-term investments.   Deficiencies are frequently met by the use
of bank overdrafts; in fact, some companies make a practice of using
overdrafts to finance stocks and debtors.   The use of cash budgets
enables shortages or deficiencies to be predicted so that suitable

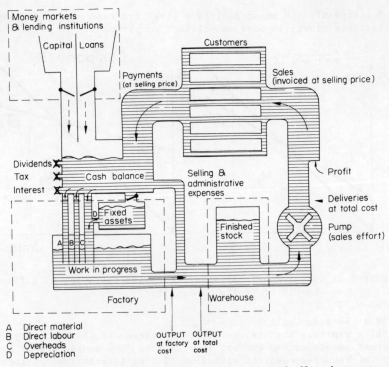

A   Direct material
B   Direct labour
C   Overheads
D   Depreciation

Figure 8.12 Diagrammatic representation of cash flow in a manu-
             facturing company

action can be taken in advance. Companies look 12 months ahead with
their cash budgets but it is advisable to take a more detailed look
at the coming three months.
     The compilation of a three-month cash budget can be seen below.

(£ thousands)		Jan.	Feb.	Mar.
Opening cash balance	(A)	100	110	110
Expected receipts				
Trade customers		80	90	90
Dividends from investments		–	–	10
Total	(B)	80	90	100
Planned payments				
Trade suppliers		40	50	50
Wages and salaries		20	25	25
Tax		–	–	55
Dividends to shareholders		–	15	–
Capital expenditure		10	–	–
Total	(C)	70	90	130
Closing cash balance (A + B – C)		110	110	80

8.10.2 Funds Flow Plan

The cash budget is concerned with money paid and received by the
company; the funds flow plan is concerned with the changes that are
likely to occur in the balance sheet. The following items are typically included in a funds flow plan covering one year.

Source of funds		(£ thousands)
Retained earnings		200
Depreciation provision		150
Increased money owing to suppliers		100
Proceeds from loan issue		500
Total	(A)	950

Use of funds		
Stock increase		100
Increased money owed by customers		150
Purchase of fixed assets		800
Total	(B)	1050

Change in cash position over the year (A – B) = –100

The funds flow plan is useful since it enables the capital requirements of the company to be planned in advance and advantage taken of
favourable conditions in the capital market, that is, when interest
rates are low or when ordinary share prices are high.

The optimum capital requirements of a company are difficult to determine; there are no standard ratios since so much depends on the
type of business, rate of growth, methods of stocking, mode of distribution, and terms of sale. If there is too little capital, profit
will be reduced and serious under-capitalisation can produce over-
trading with the danger of liquidation. Over-capitalisation is less
common and indicates inefficient use of capital.

Penalties for having too little capital are listed below.

(1) Inability to carry adequate stock thus forcing short production
runs and preventing cheap bulk-buying.

(2) Inability to cope with minor fluctuations in trade.

(3) Inability to take advantage of cash discounts or extend credit
to customers.

(4) Inability to purchase capital equipment when it is needed.

Over-capitalisation is likely to result in the following ineffic-
iencies.

(1) An inadequate return on invested capital.

(2) Excessive stocks.

(3) Loose credit policies.

(4) Poor utilisation of human and physical resources.

A company may be under-capitalised and yet either consider it inop-
portune, or be unable, to raise additional capital. In these circumstances, it is possible for the company to reduce its capital needs
so that it may continue to trade safely. The measures that can be

taken are listed below; their implementation will probably result in some loss of profit.

(1) Rent rather than buy premises; if the premises are already owned they can be sold and then leased back.

(2) Get money in more rapidly from customers by offering cash discounts or by factoring debts. Factoring is selling debts to a collecting agency which pays the company a percentage of the face value of the debt; the agency then has the task of collecting the debts from the customers.

(3) Lease or hire-purchase equipment rather than buy.

(4) Avoid paying suppliers for as long as reasonably possible.

(5) Reduce capital locked up in stock by purchasing smaller quantities of materials and parts, by reducing the batch size of parts processed, or by re-routing made-in parts to bought-out.

8.11 CAPITAL-EXPENDITURE CONTROL

Decisions on whether or not to authorise capital expenditure are among the most important faced by a company. Many capital-expenditure decisions involve a commitment to the spending of large sums of money on projects involving considerable commercial risk. Once an unwise capital investment has been made, it is often difficult to retrieve the position and there can be considerable loss of profit.

Management use yardsticks to try to assess the prospective profitability of capital-expenditure projects. The financial aspects are important but they must be considered in parallel with the technical suitability of the proposals. Some types of capital expenditure may not provide an acceptable return on the investment but may be worth while in terms of better industrial relations or greater customer satisfaction.

8.11.1 Technical Aspects of Capital-expenditure Appraisal

These call for the judgement of the engineer and cannot be quantified as can the financial aspects. The more important technical questions that should be answered when purchasing equipment are summarised below.

(1) Is the quality of work produced appropriate?

(2) How reliable will the equipment be in service?

(3) Is there adequate technical and maintenance back-up from the equipment manufacturers?

(4) Are there likely to be environmental problems?

(5) Is there likely to be shop-floor opposition to the purchase?

(6) Are there likely to be technical advances that will render the equipment prematurely obsolete?

8.11.2 Financial Aspects of Capital-expenditure Appraisal

Some types of capital expenditure, such as re-equipment of a canteen, are very difficult to appraise financially and a qualitative assessment will have to be made. However, when cash flows can be determined a quantitative appraisal may be attempted. The cash outflow is normally the installed cost of the equipment; this is likely to occur immediately and to be known accurately. Cash inflows are the estimated profits that result from the purchase. Inflows occur throughout the life of the equipment and cannot be determined with any great accuracy.

Cash flows have a time-value and some methods of appraisal take this into account by discounting. In these, future cash inflows are discounted since cash received in the future is less valuable than the same amount received now owing to the loss of interest. Similarly, cash outflows occurring at some future date are less costly than an equivalent sum of money paid out now. With a rate of interest of r (r = percentage rate/100), £1 will be worth £$(1 + r)$ after one year and, after n years, will be worth £$(1 + r)^n$. Assuming a ten per cent interest rate, £1 will grow at compound interest rates until after, say, five years it will be worth £1.61; this growth, in yearly increments, is illustrated in figure 8.13a.

The formula used to calculate the growth of value is

$$P_n = P_O (1 + r)^n$$

where P_n is the value after n years
P_O is the value at time zero
r is the rate of interest
n is the number of years

Conversely, if £1 is to be received in one year, its present value is $1/(1 + r)$ and, when the time-gap is n years, its value is $1/(1 + r)^n$. With an interest rate of ten per cent, it will be found that £1 discounted over five years has a present value of £0.621. This decrease in value is shown in figure 8.13b.

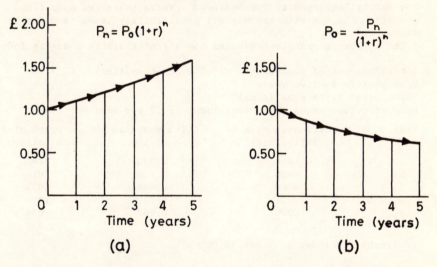

Figure 8.13 (a) Increase in value of £1 invested at ten per cent.
(b) Decrease in value of £1 discounted at ten per cent

Present value can be calculated by transposing the above formula for P_n to give

$$P_O = P_n / (1 + r)^n$$

In practice, it is more convenient to use discounting tables.
It should be appreciated that discounting is not directly concerned

with the decrease in purchasing-power resulting from inflation although high interest rates are a symptom of a high rate of inflation.

Three techniques of financial appraisal will be outlined: the pay-
back method, the net present-value method and the internal rate-of-
return method.

Pay-back Method

Here the time required for the cash inflows to pay back the cash outflows is found. For instance, if the installed cost of a machine is
£10,000 and the estimated cash inflows are £4000 each year, the pay-
back period is two-and-a-half-years (10,000/4000). A two-and-a-
half-year pay-back period is normally considered to be short enough
to justify purchase but much depends on the actual case.

Although this popular method of capital appraisal has the advantage
of simplicity, it does not take account of cash inflows after the end
of the pay-back period nor the time value of cash flows.

Net Present-value Method

In this method the cash flows during the expected life of the equipment are discounted at a given percentage rate and the total discounted cash inflows are divided by the discounted outflows to give
a profitability index. If this index is greater than unity, it means
that the annual yield on the investment will exceed the discounting
percentage. This percentage is usually chosen either as the cost of
capital expressed as an interest rate, or the rate of return required
for capital employed in the business. Where there are competing
projects, the one with the highest profitability index is normally
favoured.

The following example indicates how a profitability index is found.

Installed cost of equipment = £10,000 (cash outflow)
Machine-life is five years
Value after fifth year is nil
Rate of return required on investment is 15 per cent per annum

Year	Estimated cash inflows (£)	£1 discounted at 15%	Discounted cash inflows (£
1	4000	0.870	3480
2	4000	0.756	3024
3	4000	0.658	2632
4	3000	0.572	1716
5	2000	0.497	994
			11,846

Profitability index = 11,846/10,000
                    = 1.18

Internal Rate-of-return Method (D.C.F. Yield)

This is an alternative method of appraisal to the net present-value
method. The internal rate of return is the rate of interest, used as
a discount rate, which gives a zero value to the estimated cash
flows, that is, the discounted cash inflows equal the discounted cash
outflows. The criterion for acceptance is normally whether the
project has an internal rate of return greater than a given minimum,
such as the cost of capital.

Sensitivity Analysis

Cash-inflow estimates are based on forecasts of sales, selling prices and production costs. Sensitivity analysis enables the effect of errors in each forecast on the variability of the project to be found. With all other variables held constant, each forecast item is varied by say 10 per cent and 20 per cent on each side of its original estimated value and the effect on the profitability index or yield recalculated. When the critical variables are identi- fied, it is advisable to have the forecasts of these items carefully checked. The project could be abandoned if its success is too highly dependent on a variable whose size cannot be estimated with reason- able confidence.

Risk Analysis

This proposes a frequency distribution for each variable in the financial appraisal and from them produces a frequency distribution for the expected yield from the project. A project with a yield distribution having a narrow spread has less risk than another with a wide spread.

8.12   THE STOCK EXCHANGE

The Stock Exchange is situated in the City of London and acts as the national market for shares in public limited companies. It also buys and sells government stock and that of public bodies and local au- thorities.

8.12.1   Raising Capital

Capital is not raised directly through the Stock Exchange but through the services of issuing houses. Issuing houses act as the arrangers for a company wishing either to raise additional capital or to offer its shares to the public for the first time. Merchant banks, finance houses, and stockbrokers usually act as issuing houses. After the financial position of the company has been examined, the issuing house advises on the price at which the shares should be offered and drafts the prospectus. This is the advertisement inviting the public to subscribe to the issue. The issuing house also arranges for the appointment of an underwriter who, for a fee, agrees to purchase any unsold shares. The Companies Acts require that all shares in an issue are taken up.

Although the Stock Exchange is not the agency for raising capital, it provides a ready market for shares once they have been issued. If this assured market did not exist, investors would be discouraged from purchasing shares because of the uncertainty of being able to dispose of their investment easily and quickly.

8.12.2   Operation of the Stock Exchange

The two principal types of dealer working at the Stock Exchange are stockjobbers and stockbrokers. The jobbers act as wholesalers of shares, selling to or purchasing from brokers who act on behalf of their clients. Jobbers are approached by brokers and asked to quote a price at which they will buy and sell a particular share. The broker asks the same question of a number of jobbers, without dis- closing whether he wishes to buy or sell. He then accepts the most favourable price for his client. The difference between the selling and buying price is referred to as 'the jobber's turn'. The broker charges his client a commission for his services; he also gives free advice to clients on investment matters.

The Stock Exchange is governed by a Council elected by its members. Any member guilty of improper practices can be disciplined or suspended. A compensation fund is operated to protect clients in the event of members going bankrupt.

FURTHER READING

Carsberg, B. and Hope, A., *Business Investment Decisions under Inflation* (Institute of Chartered Accountants, London, 1976).

Goater, H.J., *Cost Control for the Manager* (Bell, London, 1975).

Hartley, W.C.F., *Introduction to Business Accounting for Managers* (Pergamon Press, Oxford, 1980).

Houghton, D. and Wallace, R.G., *Accounting Vocabulary* (Gower, Aldershot, 1980).

Rockley, L.E., *The Non-accountant's Guide to the Balance Sheet* (Business Books, London, 1979).

Shaw, J.C., *The Audit Report* (Gee, London, 1980).

Smith, J.E., *Cash Flow Management* (Woodhead Faulkener, Cambridge, 1980).

9 Design of Manufacturing Systems

9.1 THE ENGINEER AND THE MANUFACTURING SYSTEM

A manufacturing system is a complex organisation of people and equipment used to convert materials into saleable products. The conversion process employs men, machines, and energy. A manufacturing system is represented in simplified form in figure 9.1.

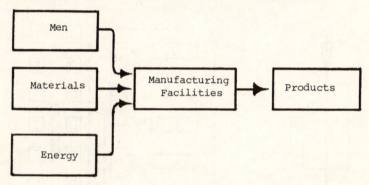

Figure 9.1 Simplified manufacturing system

Engineers are closely concerned with the design and management of manufacturing systems. The production engineer is responsible for devising a system that is able to produce the desired products at the required rate of output and at the appropriate level of quality. The production manager has the responsibility for operating the system in such a way as to ensure that output, cost, and quality-targets are met. This chapter is concerned with the design of manufacturing systems; the next discusses some aspects of their operation and control.

9.1.1 Types of Production

Manufacturing systems have to produce all types and quantities of products; hence it is hardly surprising that they vary widely in size and complexity. However, it is possible to classify engineering production broadly into flow, batch, and jobbing production.

Flow Production
This occurs where quantities are large and justify production facilities designed to produce only a single or a narrow range of products. Often special-purpose production equipment is used with either relatively unskilled operators or robots manning the work-stations. An example of flow production is the line-assembly of cars in large quantities.

153

Batch Production

Batch production can cope with the manufacture of a wide range of
work because it uses general-purpose equipment in which the tooling
is changed to suit the work being processed. The manufacturing equip
ment used in batch production has widely different rates of production; these are normally much faster than those required to satisfy
customer needs. In consequence, work is processed in batches. The
batches, which often represent many months of sales' requirements,
are then stored and used as required. Batch production is employed
in most factory machine shops and is the most common type of production. A disadvantage of batch production is the large inventory it
generates, both as work in progress and as finished stock.

Typical layouts for flow and batch production are shown in figure
9.2. The route followed by the part in the batch-production layout
will depend on which machine in each group of similar machines is
available at the time the batch is processed.

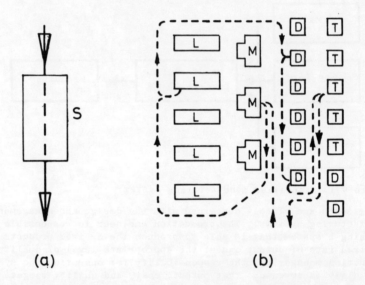

(a)   (b)

Figure 9.2   (a) Flow layout.   (b) Batch-production layout.
             S, special-purpose machine; L, lathe; M, milling machine;
             D, drill; T, tapping machine

Jobbing Production

This type of production is used when one-off-type production is required as in the manufacture of a special-purpose machine tool.
Production facilities are flexible and a high proportion of skilled
employees is required. Although normally only materials and parts
needed for the work on hand are ordered, work in progress is often
high owing to the long processing times.

9.2 PROVISION OF MANUFACTURING FACILITIES

Only an outline of this subject can be attempted in the space available. Factory buildings, plant layout, materials-handling, and
storage will be considered.

9.2.1 Design of Factory Buildings

Most factory buildings can readily be adapted to make them suitable
for the manufacture of a wide range of different products. Some
products, however, make special demands on the type of factory required; this makes the building less adaptable for other types of
production. For instance, if overhead-travelling cranes are required,
the headroom must be increased beyond the height desirable for most
purposes. Under such circumstances, it must be accepted that some
factory buildings will have limited alternative use.

For most engineering manufacture, provided that sufficient fore-
sight is used in construction, the factory can be designed for a
variety of purposes without sacrificing manufacturing efficiency.
At the planning stage the basic decision is usually whether to erect
a single-storey or multi-storey building. When land is available at
a reasonably low cost, single-storey buildings are usually preferred
for the following reasons.

(1) Greater flexibility exists for factory layout. In general
there is no limitation as to where heavy equipment can be sited and
the cost of moving plant is lower.

(2) Expansion of a single-storey building is simpler and less
costly. If no further extension on a single-storey basis is possible, mezzanine floors can sometimes be constructed or, alternatively,
part of the factory can be converted to a multi-storey building.

(3) For a given floor area, building costs are slightly less.

(4) Material-handling costs are lower.

(5) Floor vibration can be more easily localised.

(6) The risk of fire damage can be reduced.

(7) Natural lighting and ventilation can be provided through the
roof.

(8) Larger unobstructed floor areas can be used since the vertical
supports do not need to be as closely spaced as those in multi-storey
buildings.

Many companies use single-storey buildings for the factory while
housing offices and support services in multi-storey blocks.

If it is proposed to use travelling cranes or overhead conveyors,
ceiling heights of the order of 11 metres and six metres, respectively, will be required and the building itself will have to be of
stronger construction. If there is no overhead handling it is desirable to limit the ceiling height to four metres wherever possible
since this will reduce building and heating costs.

9.2.2 Plant Layout

Calculation of Equipment Requirements

Before considering detailed plant layout, it is essential to assess
the quantity and types of manufacturing equipment required to make
products in the numbers required. If established products are to be
made, it is likely that standard manufacturing times will exist for
all operations. If, however, new products are envisaged, they must
be analysed on an operation-by-operation basis as a prelude to estimating production times. Production times can be estimated using a
suitable method of work measurement (see section 10.3). The machine
running times are then found by multiplying the time estimates by
expected usage rates for each component. This calculation makes no
provision for setting time, machine breakdown, or unavoidable delays,

so the number of machine hours must be corrected by a suitable utilisation factor. In a batch-production system, the size of batches
will affect the ratio of setting to operating time and hence the
machine utilisation.

A major factor affecting equipment requirement is shift policy.
Investment in production equipment can be almost halved if two shifts
are worked instead of one.

Plant-layout Objectives
These are concerned with minimisation of manufacturing cost while
satisfying both technical and social requirements. The major objectives of plant layout are listed below.

(1) Providing the planned factory output.
(2) Reducing total manufacturing time, thereby reducing work in
progress and simplifying control.
(3) Minimising scrap caused by handling.
(4) Using minimum floor area.
(5) Ensuring safe operation of the factory.
(6) Providing high utilisation of capital equipment.
(7) Reducing total distances travelled by work in progress.
(8) Increasing operator output by installing a socially acceptable
layout and reducing fatigue.

These objectives are not all mutually compatible and the layout
selected will be generally a compromise between the conflicting requirements. When proposing a layout, it is necessary to conform with
requirements concerning space, gangways, and fire escapes. Also,
adequate provision must be made for non-productive areas such as
stores, factory offices, cloakrooms, and toilets.

Choice of Layout
In process industries, the layout is usually dictated by the manufacturing sequence. In jobbing industries, where one-off jobs are
produced, plant-layout decisions can be made in relation to past
experience and to an intuitive assessment of future needs. Where
products are made in recurring batches, demand is more predictable
and a reasonably accurate assessment of work-load can be made.

Three basic layouts are possible: process, flow, and cellular. In
mass production, quantities often justify flow production both for
component manufacture and assembly. Batch production of components
is normally associated with process layouts, although their assembly
is often on flow lines. Process layouts are usually employed for
jobbing production.

Process Layouts
These group the plant according to type; in a typical mechanical-
engineering factory the main groups may comprise CNC lathes, plug-
board capstans, presses, milling machines, and grinding machines.
Each machine will manufacture a variety of parts, the tooling being
reset between batches. These layouts are suitable where a large
variety of small and medium-sized components are manufactured in
batches. Where a number of similar machines are grouped together
there is a greater flexibility in scheduling jobs to machines and
specialisation of setting and operating skills can be practised. The

main disadvantages are due to the relatively long distances that work batches must travel between operations and the length of the total manufacturing time; these give rise to a high level of work in progress. It has been estimated that in most factories using this layout pattern, batches of work spend about 85 per cent of the manufacturing time waiting between operations.

This type of layout is sometimes called a functional layout.

Flow Layouts

These are used when a particular component or assembly, possibly with small variations, is required in numbers that justify the grouping of machines or assembly stations in operational sequence. Flow layouts are also particularly suitable for expensive components where large cost savings can be obtained by minimising work in progress. Although continuous flow is desirable, long flow lines usually have provision for inter-operational buffer stocks which help to cushion minor breakdowns. Work control is greatly facilitated with flow layouts; throughput time and hence work in progress is significantly reduced. A major disadvantage is inflexibility; another is the disorganisation resulting from breakdown of a machine in the line. On non-automatic lines there will be a loss of labour productivity since it is difficult to balance the work content at each station on the line because of variations in operator effort.

A logical development of flow layouts, used extensively for parts' production in the car industry, is to consolidate the layout into automated transfer lines operating on a fixed cycle time. Transfer lines allow the components to be passed from stage to stage automatically thereby eliminating external handling between operations and reducing direct labour cost. The high capital cost of transfer lines, and their limited flexibility, restricts their use to high-output situations where a continuing demand is expected over a number of years.

Cellular Layouts

Neither process nor flow layouts are particularly suitable for some relatively complex parts produced in small or medium-sized batches. To make these parts economically, the group-technology concept has been developed. By use of a suitable classification system, components having generally similar manufacturing operations are associated in family groups and machines are arranged in cells to enable the complete manufacture of a part in a single cell. When the operation sequence is essentially similar for all the parts in the group, a line layout within the cell is frequently adopted. However, if the sequence is likely to vary, a process layout is more convenient to permit greater flexibility in routing. The group-technology approach is not a standard procedure suitable for all applications, it simply suggests a new layout philosophy that can be developed to meet the needs of individual factories. Advantages claimed for cellular manufacture compared with process layouts are a reduction of work in progress, improved flexibility to meet changing priorities, and reduction in throughput time. It also provides greater autonomy for work groups in determining their manufacturing schedules, a greater sense of involvement, and the possibility of job enrichment by rotation of jobs and acquisition of setting skills. The main disadvantage is the low utilisation of some machines in the cell, particu-

larly when a line layout is used. In practice, job mobility of
operators may be difficult to achieve owing to union agreements and
the cost of multi-skill training.

Work Flow

A number of algorithms have been proposed to assist in obtaining the
best process layout. They are based generally on minimum-travel considerations or adjacency-preference ratings. The practical usefulness of these algorithms is strictly limited.

   The plan shape of the factory and access points for loading and
unloading impose constraints on the layout and siting of material
stores or warehouses. Within these constraints the layout should
allow, as far as possible, for uni-directional work flow with a minimum of back-tracking. Decisions to facilitate uni-directional flow
can be made only by comparing the operation sequences for a large
number of the parts manufactured. On the basis of experience it is
known that some processes are used predominately for first operations
and others for subsequent operations. As an example, in mechanical-
engineering production, the sequence of manufacture is likely to be
as shown in figure 9.3.

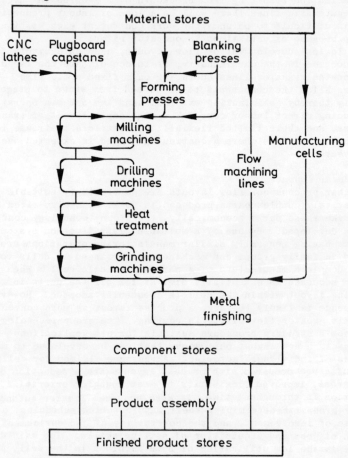

Figure 9.3   Typical material flow pattern in engineering manufacture

When there is no clear-cut operation sequence on which to base a layout, travel charts can be used to show the pattern of work in a factory and to provide a guide to plant layout; a typical example is shown in figure 9.4. The starting-points for batches of work are listed vertically and their destinations are listed horizontally; the figures in the intersecting squares represent the number of batches passing between any two departments. Numbers appearing below the diagonal line indicate reverse flows, which would occur if the departments were situated in the listed sequence, whereas the scatter of points above the line indicates the variety of operation sequences experienced. Travel charts are time-consuming to construct if there is a large number of different components being processed and it may be necessary to base the chart only on a representative selection of parts.

Assembly Organisation

Depending on the assembly processes required, the organisation of assembly can be based on individuals, groups, or lines. When quantities are very large, automatic assembly may be possible. In many companies where large batches of similar or identical products are

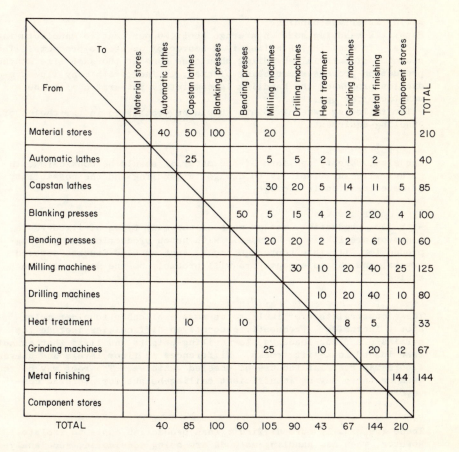

From \ To	Material stores	Automatic lathes	Capstan lathes	Blanking presses	Bending presses	Milling machines	Drilling machines	Heat treatment	Grinding machines	Metal finishing	Component stores	TOTAL
Material stores		40	50	100		20						210
Automatic lathes			25			5	5	2	1	2		40
Capstan lathes						30	20	5	14	11	5	85
Blanking presses					50	5	15	4	2	20	4	100
Bending presses						20	20	2	2	6	10	60
Milling machines							30	10	20	40	25	125
Drilling machines								10	20	40	10	80
Heat treatment		10		10					8	5		33
Grinding machines						25		10		20	12	67
Metal finishing											144	144
Component stores												
TOTAL		40	85	100	60	105	90	43	67	144	210	

Figure 9.4 Travel chart

made, assembly lines offer the advantages of rapid throughput and
minimum floor space. The social disadvantages of line assembly are
considerable when operators are physically separated and cannot
easily talk to each other, as on car assembly lines.

For jobbing and small-batch assembly there are social advantages in
using groups of operators who mutually agree their work arrangements.
Such an approach encourages versatility and the formation of coherent
working groups. It is essential that provision be made for operator-
training and group leaders must be carefully selected.

The Integrated Layout

Having decided on the floor area required for each section and the
preferred sequence of sections, the situation of each can be indicated on a floor plan of the factory, making provision for gangways,
materials-handling equipment, work in progress, and factory services.
A more detailed plan can then be attempted using three-dimensional
models or scaled templates of production equipment to establish the
positions of individual items. A preliminary layout of the factory
is thus provided as a basis for discussion before a final layout is
agreed.

9.2.3 Materials-handling

Materials-handling adds on average about 20 per cent to manufacturing
costs. In addition, handling is a source of damage to products. The
planning of materials-handling should be done at the same time as the
production processes are being planned; in fact, with many flow production processes the handling is completely integrated with the
manufacturing operations.

The following general points should be considered in the choice of
handling equipment.

Material to be Handled

The size, weight, and form of the material should be considered, together with any liability to damage in handling and the possibility
of deterioration in storage.

Type of Production

With flow production, one or more of the many types of fixed-route
conveyor are likely to be used. With batch-production and jobbing-
production, flexible route equipment, such as hand trucks, powered
trucks, and portable conveyors will probably be the most suitable
choice.

Limitations Imposed by Buildings

These constraints are likely to be more applicable to older multi-
storey factories. Frequently encountered difficulties include low
permissible floor-loadings, low ceiling-heights and lifts that cannot
take loaded fork-lift trucks. Differences in floor levels and uneven
floors may restrict the use of wheeled vehicles. The roof may be too
weak or there may be insufficient ceiling-height for overhead conveyors.

Handling-costs

These are usually more difficult than production costs to isolate.
However, when new handling-methods are being considered, some analysis of the anticipated cost benefits should be attempted.

9.2.4 Storage

Owing to the variability of the shape and type of material stored, there is no single preferable method of storage. However, economy of space and handling-effort should always be in mind when planning storage facilities.

To facilitate handling in unit loads, items for storage should be palletised whenever possible. The storage of pallets in racks is preferable to the stacking of pallets on each other since the lower pallets can be withdrawn from racks without moving those stored above.

To maximise space-utilisation in palletised stores, stacker or reach-type fork-lift trucks should be used. These trucks enable aisle widths of two metres to be used as compared with three metres for a conventional fork-lift truck.

Small parts stored in bulk can be placed in standard containers that fit into specially designed storage racks. Where possible, the same container that has been used to hold parts during manufacture should be used to hold them in storage; this reduces handling-costs and minimises damage.

Goods can be stored on or in a variety of pallets. A common type is the wooden pallet base; here the load is secured on the pallet. These loads can be stacked provided that they are sufficiently rigid and can be built into box-like shapes. There are also box-pallets and post-pallets which are rigid and stackable irrespective of their contents. When specifying pallets, British Standard sizes should be used since they are designed to provide exact fits for pallets into containers.

9.3 DESIGN OF WORK SYSTEMS

By using method study and ergonomics, it is possible to design new systems or to improve those that are already in use. Ergonomics uses physiological and anatomical data to improve work-place design and is complementary to method study. Space does not permit a discussion of ergonomics despite its interest; the rest of this chapter deals with the older-established technique of method study.

Method study is that part of work study that provides a systematic approach to improving the way in which work is done. It offers significant increases in productivity at little expenditure since it usually operates within the existing product design and capital equipment.

The procedure of method study has been formalised into six steps as listed below.

(1) Select work to be studied.
(2) Record existing method of working.
(3) Examine critically the existing method.
(4) Develop an improved method.
(5) Install the improved method.
(6) Maintain the improved method.

Method study is often viewed by workers and supervisors as a threat to their traditional methods of working and to their security of employment. Unless these fears can be allayed, resistance will be offered which could, in extreme cases, prevent the application of method study. However, if there is a sympathetic understanding of the fears of those affected, co-operation should be obtainable.

Trade unions will generally accept work study at national level but
it would be unwise to assume similar acceptance at plant level. Before method study itself is discussed, ways of securing employee
co-operation are outlined.

9.3.1 Human Factors

Possible fears of workers and relevant management action are summarised below.

Redundancy

Here a guarantee is required that no person will lose his job as a
direct result of method study. This should not be difficult to give,
provided that the level of labour turnover is sufficient to absorb
any surplus labour created by improved methods.

Loss of Earnings

The company should be able to promise that earnings will not be reduced.

De-skilling

There may be simplification of work and it will be difficult to give
guarantees here.

Break up of Working-groups

Some regrouping may be necessary but existing groups should be retained whenever possible.

Working Harder

It should be explained that the object of method study is to design
a more efficient work system and that any previously heavy physical
work is likely to be eliminated.

   A powerful and effective argument for method study in the private
sector is that unless the company improves its efficiency, relative
to its national and international competitors, sales are likely to
fall and total redundancy may result.

9.3.2 Selection of Work to be Studied

It is hoped that senior management will influence the choice of major
method-study projects to be undertaken since they are in the best
position to decide on priorities. The following factors, either
jointly or separately, indicate possible candidates for investigation
by method study.

   (1) A high proportion of the job cost is made up of direct labour
cost.
   (2) Work on which target production costs are not being met.
   (3) Production bottle-necks.
   (4) Poor space-utilisation.
   (5) High handling-costs.
   (6) Dirty or dangerous work.
   (7) Low direct-labour utilisation.
   (8) High reject-rates.

   A check should be made on the present and future rate of production
since this will relate directly to the total potential savings. It
is probably wise to avoid work where labour trouble is simmering

since the investigation may cause the situation to boil over. The
techniques that are expected to be used and the objectives of the
investigation should be selected at this stage.

9.3.3 Recording of Existing Method

The facts concerning the present method should be recorded from
observations in the factory rather than from records. A large
variety of charts is available augmented, where appropriate, by
photographic techniques.

The symbols in general charting use are shown below.

○ operation

□ inspection

◁ transport

▽ storage

◗ delay

The various recording charts can be grouped into the following types.

Sequence-type Charts

These are either outline or flow process charts. Outline process
charts use only operation and inspection symbols whereas flow process
charts use all five symbols. The subject of both charts can be
either men or materials. A materials-type outline process chart is
shown in figure 9.5. By comparison, figure 9.6 is a flow process
chart for operations 5 and 6 of the outline process chart in figure
9.5; it will be seen that the number of stages has now increased from
two to eight.

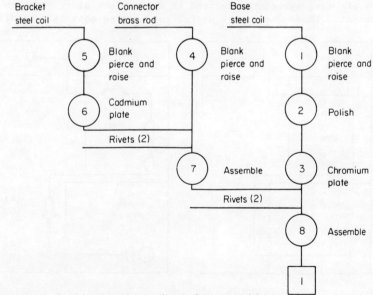

Figure 9.5 Outline process chart for assembly

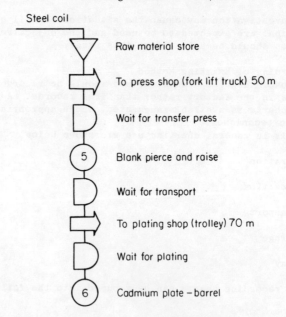

Figure 9.6 Flow process chart for operations 5 and 6 in figure 9.5

Layout-type Charts
These indicate movement of material or men. The simplest is a string
diagram; this consists of a scale plan of the working area in which
string is used to show the movement path. String diagrams are of
limited value but can be useful when considering stores and other
layout problems. The flow diagram provides greater detail with the
five charting symbols being used as appropriate along the path of
movement. These charts are useful for solving materials-handling

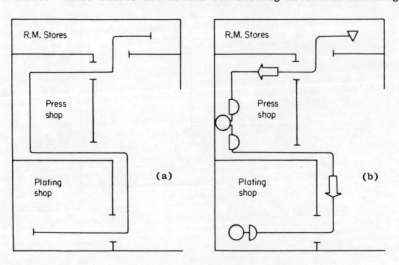

Figure 9.7 Layout-type charts: (a) string diagram, (b) flow diagram

problems and minimising operator movement. Both types of layout
chart are shown in figure 9.7.

Time-based Charts

The principal chart here is the multiple-activity chart. This chart
enables the working-waiting pattern of a group of men and machines to
be established. As will be seen from figure 9.8, a separate column
is allocated to each man or machine and their activities are shown
against a time-scale. Another chart drawn against a time-scale is a
simo chart, which is described in the next section.

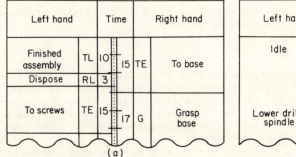

Figure 9.8 Multiple-activity chart showing the effect of second
machine operator in improving machine utilisation

Two-handed Charts

These charts are used to record in detail the movement of each hand
in a single operation. The simo chart is drawn against a time-scale
based on the wink (1 wink = 1/2000 min. $\approx \frac{1}{2}$ frame of ciné film shot
at 16 frames per second). The two-handed process chart uses the flow
process chart symbols to show in sequence the work of each hand.
Both types of two-handed chart are used for assembly operations;
examples of each chart are shown in figure 9.9.

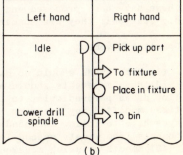

Figure 9.9 Two-handed charts: (a) simo charts, (b) two-handed
process chart

9.3.4 Examination of Existing Method

After the facts concerning the original method have been accurately presented on the most appropriate type of chart, they should be examined critically so that ideas can be generated for an improved method. A systematic approach favoured in the United States is to use a check-list designed for the particular type of chart employed at the recording stage. A more general approach is to complete a critical examination sheet for each of the key operations that have been recorded. An example of a critical examination sheet is shown in figure 9.10. At this stage, all ideas are noted for subsequent sifting with the investigator normally working alone.

PRESENT FACTS		ALTERNATIVES	ACTION
what is done	why	what else	what should be done
how it is done	why that way	how else	how it should be done
when it is done	why then	when else	when it should be done
where it is done	why there	where else	where it should be done
who does it	why that person	who else	who should do it

Figure 9.10 Critical examination sheet

The investigator should work from a consideration of the whole process and not involve himself at an early stage in improving single operations. The ideas generated will probably fall into three categories.

(1) Elimination of unnecessary work.
(2) Simplification of work.
(3) Improvement in the sequence in which work is done.

Principles of Motion Economy

These have been proposed and provide a guide to the design of assembly stations; they are summarised below.

Operator Movement

(1) Both hands should do useful work except during rest periods.
(2) Hand movements should be simultaneous and symmetrical about a line drawn perpendicular to the operator's chest. This rule is sometimes modified, by ignoring the symmetrical requirement, if the eyes have to follow the hands as with the picking-up of small components from widely spaced locations.
(3) Smooth swinging movements are to be preferred since movements that involve sharp directional changes take longer and are more tiring.

(4) Use other parts of the body, for example the feet, to assist the hands.

(5) Employ momentum and gravity; this is sometimes possible in the disposal of work.

Work-place Design

(1) Tools and materials should have fixed locations to ensure that each job has a constant movement pattern.

(2) Power-assisted tools and work-handling equipment should be used where appropriate.

(3) Benches of correct height should be used; a height of 970 mm (38 in) will permit both sitting and standing work.

(4) Chairs should match the bench height and have adjustable seat-height and back-rest to suit individual operators.

(5) Where a high degree of muscular control has to be exercised, as with some fine assembly work, arm rests should be provided.

(6) The layout should minimise the hand and body movements of the operator.

An example of an assembly layout incorporating a number of the above points is illustrated in figure 9.11.

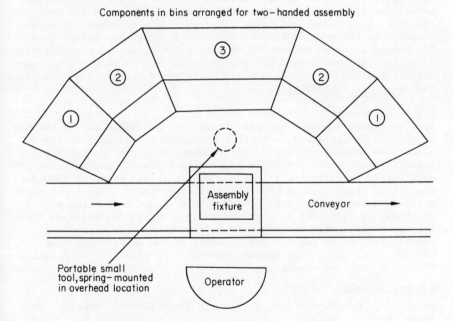

Components in bins arranged for two-handed assembly

Figure 9.11   Plan view of assembly layout

9.3.5   Development of Improved Method

Here the ideas from the previous stage are shaped into a firm proposal for an improvement in the working method. This will probably involve discussions with interested parties and, possibly, modification of the original ideas as a result of these discussions.

Charts are prepared showing the proposed method and the expected savings are estimated.

On larger projects, a formal report may be required at this stage. The report should preferably be brief and include the following information.

(1) Summary of recommendations
(2) Estimated savings
(3) Capital expenditure
(4) Time required for implementation
(5) Possible difficulties
(6) Acknowledgements of assistance

Supporting arguments, charts, and detailed costings should be relegated to appendices.

9.3.6 Implementation of Improved Method

In simple instances, the investigator can deal with the implementation stage himself. However, in major projects it is preferable that a senior member of management be given general responsibility. A works manager, for example, is likely to be more effective than a member of the work study department in ensuring that an unwilling supervisor or section manager meets a difficult deadline. The co-ordination of major projects should, however, remain in the hands of the work study investigator who can draw up any necessary network diagrams of the work to be done and monitor progress against them. The investigator should be active during the critical first few days after the improved method has been introduced. He should encourage where necessary and resist hastily contrived modifications to overcome teething difficulties.

9.3.7 Maintenance of Improved Method

Once an improved method has been installed, it should not be changed unless the change brings a further improvement. In most instances, authorised methods of working are recorded in operation layouts; these form the basis for works orders and standard costing data. The work method is described in greater detail in the standard time documentation maintained by the work study department. Since documentation is amended only after an authorised change in working method, it should be difficult in a well-organised and well-supervised factory to deviate from the improved method. Changes to authorised methods that bring genuine improvements should be welcomed from wherever they come; suggestions from the shop-floor can be formalised through suggestion schemes.

FURTHER READING

Drury, J., *Factory Planning Design and Modernisation* (Architectural Press, London, 1981).
Morrison, A., *Storage and Control of Stock* (Pitman, London, 1981).
Pemberton, A.W., *Plant Layout and Materials Handling* (Macmillan, London, 1974).
Ranson, G.M., *Group Technology* (McGraw-Hill (UK), Maidenhead, 1972).
Whitmore, D.A., *Work Study and Related Management Services* (Heinemann, London, 1976).

10 Operation of Manufacturing Systems

In the previous chapter various aspects of manufacturing system design were outlined. Here the operation and control of a manufacturing system is considered under the headings of quality control, materials management, labour control, project management, maintenance management, and management information systems.

10.1 QUALITY CONTROL

The design department of an engineering company indicates the specified level of product quality by means of the drawings and specifications it issues. The quality levels it specifies should be appropriate to the company's products and to its markets. The marketing and production departments are also concerned with quality standards but frequently have differing views on the level that is appropriate.

Comparison of actual and specified levels of quality is the responsibility of the manager of the quality control department. It is desirable that he should report to the design director, rather than to the production director; this arrangement reduces the risk of undue pressure on quality control staff to pass substandard work.

10.1.1 Measurement of Quality Performance

There is no single parameter that enables an adequate assessment to be made of product quality. However, three factors taken together provide management with a reasonable idea of the general level of achievement.

(1) The value of scrapped work.
(2) The cost of rectifying faulty work.
(3) The quantity of faulty products returned under guarantee by customers.

These factors can be measured monthly and plotted to establish trends. To avoid adjustments for inflation and/or variations in output, items (1) and (2) can be expressed as a percentage of the value of monthly production and item (3) can be expressed as a percentage of monthly output.

10.1.2 The Organisation of Inspection

Manufacturing operations always produce a proportion of substandard work. It is the task of the quality control department to monitor actual quality levels and to reject work that does not reach the specified level of quality.

Purchased supplies, such as raw materials and components, are normally sampled and checked before they are accepted into stock. If the level of quality is below that specified, the whole consignment

of goods may be returned to the supplier. However, should supplies
be needed urgently, the goods are sorted and only those that are
faulty are returned.

Factory-produced parts and sub-assemblies are normally monitored
for quality during production and substandard work is rejected. It
is desirable that faulty work is detected as soon as possible after
it has been produced; the two control charts described in the next
section are suitable for this purpose.

It is usual for finished products to be given a final inspection
which can take the form of functional and endurance tests. Some
companies undertake an inspection audit after final inspection; this
is a detailed reinspection of a small sample of goods awaiting despatch to customers. The audit may be undertaken by the quality
control department in some companies or by the design department in
others. Likewise, the analysis of faults in products returned under
guarantee may or may not be the responsibility of the quality control
department.

10.1.3 Statistical Quality Control

The inspection of every item produced or received is not only expensive but is less accurate than might be expected. This inaccuracy
is due to the boredom created by 100 per cent inspection and the con-
sequent careless mistakes of inspection staff. Sampling plans based
on statistical theory not only reduce inspection costs but still
provide a reasonably representative impression of total quality.
Sampling techniques used to control production fall into two distinct
types. One is where there is an actual measurement of a variable
quantity, such as size or weight, and the other is where an attribute
is checked on an accept or reject basis, for instance, when using a
'go' and 'not go' gauge to check a part.

Control Charts for Variables

Mean and range charts are used to control the size of a particular
variable such as the diameter of a shaft. Examples of mean and range
charts are shown in figure 10.1. Samples, typically of four to six
parts, are taken from production and measured at given intervals of
time or output. In the case of the shaft diameter, the average
diameter of the sample is plotted on the mean chart and the difference between the largest and smallest shaft is plotted on the range
chart. The process itself cannot be said to be in control unless
both mean and range charts are in control, that is, the plotted
points lie between the limit lines on the charts. It is of interest
to note that although the production process may produce parts that
form a skewed or multi-modal distribution, the distribution of sample
means will form a normal distribution.

Control chart limits are conventionally fixed so that, with the
process in control, the warning limits will exclude 5 per cent of the
samples and the action limits will exclude 0.2 per cent of the
samples. The warning and action limits are drawn respectively at
1.96 and 3.09 standard deviations from the chart means. Usually,
upper warning and action limits only are drawn for range charts;
the lower limits are close to zero and of little interest in engineering manufacture. Should a point on either the mean or the range
chart fall between the action and warning limits, a second sample is
taken. If this second sample falls inside the warning limits, the
process is allowed to continue. Should two successive samples fall

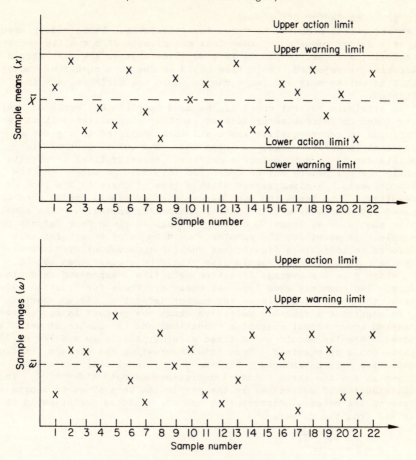

Figure 10.1   Mean and range charts for samples

between action and warning limits, the process is stopped and necessary adjustments made. The process is also stopped if a point on
either chart falls outside the action limits. Warning limits are
sometimes omitted completely from control charts since some quality
engineers consider them to be an unnecessary complication.
Control chart limits are constructed by taking about 25 samples
from the running process and calculating the average size, $\bar{X}$, and the
average range, $\bar{\omega}$, of the samples. As the average sample range for
any given sample size is directly proportional to the standard deviation, control chart limits can be established by multiplying $\bar{\omega}$ by
appropriate constants. Tables of constants for mean and range charts
may be found in BS 2564.
The limits of accuracy of the production process itself may differ
considerably from the limits of accuracy that the designer has specified on the drawing. If the process has wider limits, an attempt
must be made to find a more accurate process. Should process limits
be much narrower than the designer's limits a less accurate and less
expensive manufacturing process should be sought.

Control Charts for Attributes
Control charts also can be drawn for attributes. These widely used
charts do not require any numerical measurement of a quality characteristic; all that is needed is a decision whether the part should be
accepted or rejected. Rejection could be due to a number of causes;
this is unlike mean and range charts where one variable only is
charted.

An attribute control chart can be used to plot the levels of quality found in successive production samples; alternatively it may be
employed to record quality where all work completed in a given time,
for example daily output, is inspected. The out-of-control high
points can be used to detect significant deteriorations in quality
while the low points may be converted into more permanent quality
improvements. A disadvantage of this type of chart is its lack of
sensitivity and hence the much larger size of sample needed when compared with those used with control charts for variables. The sample
size should be at least 50 or large enough to yield four defects per
sample. If possible, the samples should be large enough for zero
defects to indicate a significant quality improvement over standard;
for this to occur, the sample size should be greater than $(9 - 0.9\bar{p})/$
$\bar{p}$, where $\bar{p}$ is the average 'fraction defective', expressed in decimal
form. Two commonly used types of chart are those for fraction defective, the p chart, and those for number defective, the np chart.

To construct a fraction defective chart the process is successively
sampled under normal operating conditions and the number of defects
noted. Samples should be of fixed size; typically 20 samples of 100
parts could be examined. From this information the average fraction
defective $\bar{p}$ is calculated. Warning and action limits can now be
drawn at two and three standard deviations on both sides of $\bar{p}$. The
distribution of defectives in the sample can be assumed to approximate to the binomial distribution and the standard deviation is obtained from the formula

$$\sigma = \sqrt{\frac{\bar{p}\ (1 - \bar{p})}{n}}$$

where σ is the standard deviation of a binomial distribution
      n is the sample size.

Number defective charts are similar in principle to fraction defective charts. They are slightly simpler to understand since they
record the number of defectives in each sample rather than the fraction defective. The method of construction is similar to that described above except that the standard deviation is found from the
following expression

$$\sigma = \sqrt{n\bar{p}\ (1 - \bar{p})}$$

An example of a number defective control chart is shown in figure
10.2.

Sampling Plans for Incoming Parts
Here the inspection is also by attributes and a variety of plans can
be used.

Single sampling simply accepts or rejects the consignment based on
a single sample. A sample size n is determined; then, should the
number of rejects in the sample exceed a given limit c, the whole

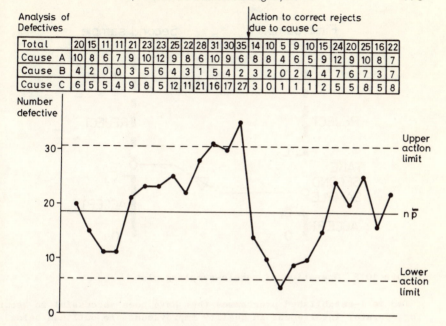

Figure 10.2   Number defective chart

batch of work is returned to the supplier. With this type of plan
the sample is often several hundred parts.

Inspection costs can be reduced if double-sampling plans are used
since the sample sizes are smaller. If the number of rejects in the
first sample n_1 is equal to or less than the acceptance limit c_1, the
batch is accepted; if the number of rejects in the first sample is
greater than the rejection limit c_2, the batch is rejected. Should
the number of rejects in the first sample be greater than c_1 but less
than or equal to c_2, a second sample of larger size n_2 is taken and
inspected. The plan then proceeds as a single-sampling scheme with
a sample size of $n_1 + n_2$. If the sum of the rejects in both samples
exceeds a new rejection limit c_3, the batch is rejected; if it is
less than or equal to c_3, the batch is accepted. The operation of a
double-sampling scheme is shown in figure 10.3.

Sampling plans to suit particular applications can be designed by
consulting either the DEF Tables or the Dodge and Ronig Tables,
details of which are given at the end of this chapter.

10.1.4 Quality and People

Although the inspection department can monitor quality standards,
good quality comes largely from the care and concern of those directly involved in manufacture.

There are many ways of directing the attention of production operators to improving the quality of the work they produce. One method
is to introduce a wage-payment incentive scheme in which one of the
elements is the quality of work produced. Another is to install at
each work-station a small frame carrying a card containing a list of
quality points relating to the work being done.

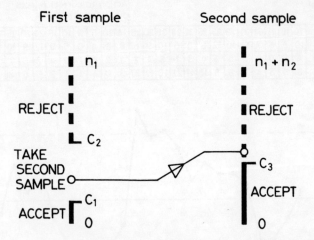

Figure 10.3 Operation of double-sampling plan

   Two well-established programmes that have been successful in securing operator involvement in quality improvement are outlined below.

Zero-defects Programme
This plan which attempts to involve operators in product quality was
originated in the United States. Operators pledge themselves to
improve quality levels by avoiding errors in their work; they also
are asked to inform their supervisors of the causes of work errors
and to suggest ways in which the errors can be removed.

Quality Circles
These were established in Japan in the early 1960s and now have
world-wide application. The circles are led by the production foreman and consist of up to a dozen operators. The group itself usually
selects the quality problems to be discussed although it may be asked
to examine problems by management or by another department. A typical topic discussed is the reasons for scrapping of expensive parts
and how this loss could be reduced. A Pareto analysis of the distribution of scrap or rectification cost is often used to direct
circles to the areas of greatest loss. Fishbone diagrams are helpful
in the analysis of causes and sub-causes for reject work. An example
of a fishbone diagram is shown in figure 10.4.
   It is usual for a number of quality circles to operate simultan-
eously in a single factory. The person appointed by management to
assist with the organisation of quality circles is referred to as a
facilitator.

10.2 MATERIALS MANAGEMENT
Materials management provides a production plan that is consistent
with sales and manufacturing capacity; it ensures that materials and
parts are obtained at the appropriate time and controls output in

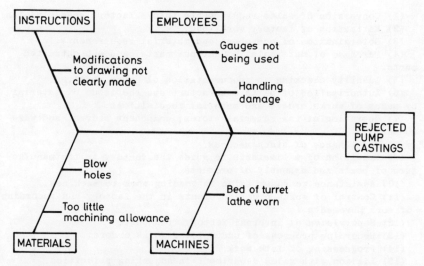

Figure 10.4 Fishbone diagram

conformity with the plan. The organisation of this function is sometimes the responsibility of one department although a more traditional approach is to divide the responsibility between a production
control department and a purchasing department. When organised as
two separate departments, production control is generally subordin-
ated to the works manager and purchasing to the financial director.
With a single department it is more likely that the materials manager
will be answerable to the production director. The basic communications network for materials management is shown in figure 10.5 and
its range of activities are as follows.

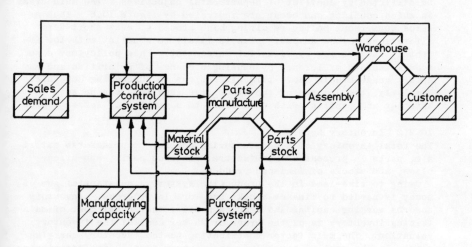

Figure 10.5 Materials management communications network

(1) Conversion of sales requirements into a factory assembly plan.

(2) Estimation of factory work-load.

(3) Determination of component and material requirements.

(4) Purchase of materials, bought-out parts, and subcontracted parts.

(5) Quantity checking and documentation of goods received.

(6) Authorisation of parts manufacture and the issue of material by means of works orders and material requisitions.

(7) Operation of raw material stores, component stores, and warehouse.

(8) Maintenance of stock records.

(9) Provision of a timetable to guide the factory in the manufacture of parts and assembly of products.

(10) Assistance to supervision in loading work to machines.

(11) Control of works order documents in the factory and recording of work movements.

(12) Supervision of internal factory transport.

(13) Recording progress of uncompleted works orders.

(14) Progressing of late work.

(15) Liaison with sales department in adjusting priorities.

(16) Despatch of finished products to customers.

(17) Implementation of product-design changes.

In most factories some of the responsibilities listed will be apportioned to other departments; for example, the sales department is likely to assume responsibility for the finished-product warehouse and for despatch to customers. This sort of curtailment of its scope is unlikely to affect the operating efficiency of materials management. However, a restriction such as separate control of stores and stock records is likely to prove unsatisfactory.

10.2.1 Relationship with Other Departments
Production control provides an excellent example of how a company can be afflicted by conflict of departmental objectives. The main areas in which conflict can occur are indicated by figure 10.6. Thus a sales department policy requiring high stocks to meet short-term alterations to the assembly plan is likely to conflict with low inventory targets set by financial management. This policy may also reduce factory efficiency by causing a change of priorities and may entail unacceptable short-term alterations to purchasing schedules. Similarly, the demands of the other five functions can be seen to interact and conflict with each other to a greater or lesser degree.

10.2.2 Inventory Levels
The total inventory of a manufacturing company comprises raw materials, parts in process of manufacture, finished parts, sub-assemblies, and stocks of finished products.

Owing to time-lags in the production system, a substantial sum of money is needed to finance stocks and work in progress. Inventory absorbs working capital and reduces profit; the annual cost of maintaining inventory is of the order of 25 per cent of the inventory valuation. The main factor contributing to the inventory-carrying charge is the cost of borrowing money; other factors include storage costs and losses due to deterioration and obsolescence.

The amount of capital tied up in inventory will depend largely on the type of production used in the factory. If flow production is

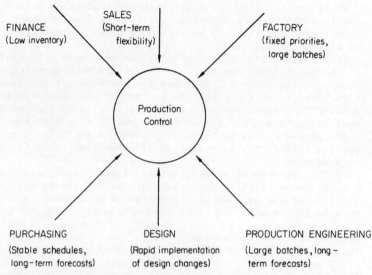

Figure 10.6 Possible areas of conflict with production control

used, inventory will be low; when batch production is used, inventory will be greater although here the size of batch chosen will exercise a considerable influence on the level of inventory. Should the company adopt a sales policy of satisfying widely varying customer demand from warehouse stock, a large inventory of finished products will be needed. Another factor affecting inventory is the availability of material and bought-out parts. If purchasing lead-times are subject to large and rapid fluctuations or if suppliers are affected by strikes, companies will seek to protect themselves by carrying high stocks.

10.2.3 Capacity
Factory capacity should be known by those planning production. In the case of factories having flow-production layouts, the rate of output is relatively inflexible and capacity is likely to be close to the designed production rate of the plant. There is greater flexibility with batch-production and jobbing-production layouts and capacity is usually found by reference to past achievement. Some companies refuse to state a capacity ceiling for their factories since they find that there is continuing elasticity.

Factory capacity can be increased by the following measures

(1) Working overtime and additional shifts
(2) Increasing the labour force
(3) Installing additional equipment
(4) Installing more productive equipment
(5) Improving direct labour utilisation and performance
(6) Improving plant utilisation

The output of finished products can be increased by subcontracting parts' manufacture and raising the proportion of parts purchased from outside suppliers.

It is particularly important that the production plan should not overload the factory. Serious overloading creates chaos on the shop-floor, produces high and unbalanced stocks, and reduces the output of finished goods.

If the work-load falls, remedial measures should be taken promptly. In the short term, overtime can be cut and subcontracted work brought back into the factory. If these two remedies are not available, the total labour force may have to be reduced and consideration given to selling some fixed assets. If manufacture is by flow production, the most economic method of reducing output is to reduce the number of hours worked. This is because flow-production lines cannot usually operate efficiently at output rates other than those for which they were designed.

If the products being manufactured have a pronounced seasonal demand, two strategies are possible. Either stocks will have to be built up during periods of low demand or production capacity will have to be expanded and contracted seasonally to match demand.

10.2.4 Periodic Reordering
The two major methods of parts-ordering are periodic reordering and stock-level reordering. Periodic reordering is suitable for both flow-production and batch-production factories. Period batch order-ing is a type of periodic reordering; it is described below.

A production plan is prepared periodically, often monthly. This is then exploded by use of parts lists to indicate the parts require-ments. Parts requirements are next reduced by any surplus stocks of parts likely to be left over from previous production plans and additions are made for anticipated scrap. These adjusted require-ments become the order quantities although some may be increased if any are considered too low to be manufactured economically. If sub-assemblies are stocked, rather than being made at the same time as the main assembly, the ordering procedure is complicated since sub-assembly stocks will also have to be taken into account.

Order must be issued to the factory with the correct lead-time if parts are to be available for assembly when required. Figure 10.7 shows the scheduling of a simple product having a multi-stage assem-bly. In batch production it is generally accepted that batches of parts are being worked on for about 15 per cent of the time that they are in the factory; during the rest of the time they are waiting for a machine, an operator, inspection or transport to the next stage of production. As a rule of thumb, many companies allow a process time of one week per operation when scheduling batch pro-duction.

Periodic reordering produces lower inventory levels than stock-level reordering but demands a considerable clerical effort. This clerical work has been greatly expedited by the use of computers. Indeed, comprehensive computer-aided production control systems are available. Some are modular, allowing modules such as plant-loading to be added to a core that would consist of the basic ordering system.

10.2.5 Stock-level Reordering
This method of ordering is simple, flexible, and widely used; its operation can be seen by reference to figure 10.8. Here is a graph of stores stock drawn against time for a part ordered by a stock-level reordering system; the graph has been simplified by assuming

that the rate of usage is constant and that the order arrives into stock in a single lot.

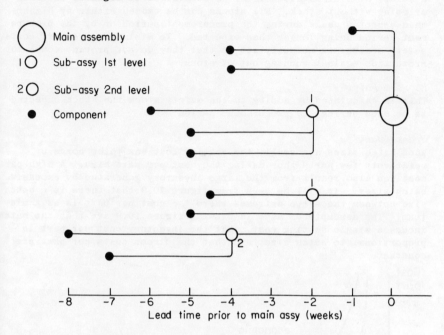

Figure 10.7   Manufacturing schedule for multi-stage assembly

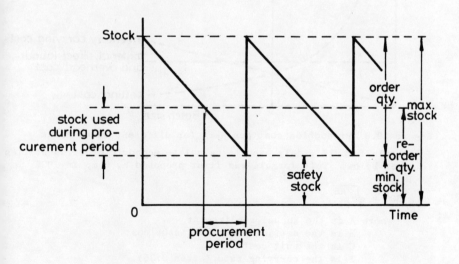

Figure 10.8   Stock/time graph

Safety Stock

A safety or minimum stock level is maintained to minimise the risk of being without stock. Nil stocks can be caused either by higher-than-expected usage during the procurement period or by the procurement period being longer than expected. To minimise inventory costs, safety stocks are normally set so that they do not provide complete protection against running out of stock.

Reorder Point

This is calculated by adding to the safety stock the stock expected to be used during the procurement period.

Order Quantity

When batch sizes are small, the setting cost has to be borne by relatively few parts thus making the cost per part high. A high part cost can also result from the large inventory generated by excessive batch sizes. It will be seen from figure 10.9 that there is a batch size between these two extremes where the cost per part is at a minimum. The assumptions made in drawing figure 10.9 are that the batch incurs a single setting cost, that the inventory cost per part is proportional to batch size, and that the direct costs per part are constant.

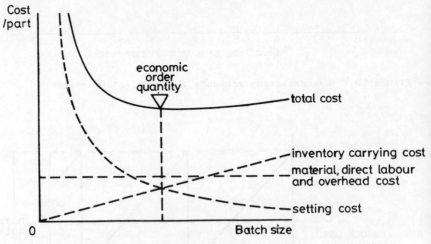

Figure 10.9 Production cost per part for different batch sizes

A simple formula widely used to calculate economic order quantities is given below. This formula was first proposed by Camp in 1922.

$$EOQ = \sqrt{\frac{2AS}{CI}}$$

where A is the annual requirement
S is the setting and order issue cost
C is the unit cost
I is the carrying rate (often 0.25).

It should be noted that the cost per part is not very sensitive to batch size in the region of the most economic batch quantity. Very small batches should, however, be avoided because of their high cost per part.

Average Stock
The average level of stores stock for a part ordered by this method
is half-way between its maximum and minimum levels. The total expected stores inventory for a stock-level reordering system is the
sum of the average stocks of all parts carried in stores.

Ordering
There are two basic monitoring systems in common use for deciding
when another order for a part should be issued.
    The first uses a record that maintains two current balances, one of
stores stock and the other of outstanding orders. When the sum of
stock and outstanding orders falls to below the reorder point, a new
order is issued. These records can be kept either on record cards or
by a computer. Owing to the reducing cost of computers and the
availability of a whole range of suitable software, computerised
systems are extensively used. Not only will the computer indicate
when a new order is required, it will also calculate the economic
order qualities and update the usage rates on which reorder levels
and order sizes are based.
    The second ordering system is suitable only for parts where the
procurement time is much less than the period between orders. In
this system the reorder quantity is physically segregated from the
rest of the parts in stock; for instance, it could be placed in a
heat-sealed polythene bag together with a reorder card. When no more
loose parts are available for issue from stock, the seal on the bag
is broken and the reorder card processed to raise a replacement
order. Sealed-stock reordering does not require a stock record to be
maintained.

10.2.6 Ordering Based on Cost Classification
The exclusive use of a single method of ordering is not always desirable; a popular multi-mode method is based on the cost classification
of parts. Most engineering products, when analysed by cost, follow a
Pareto distribution similar to that shown in figure 10.10.
    The expensive class A parts can be ordered in small quantities,
preferably using a periodic reordering system; this will minimise
inventory cost, particularly if they can be speeded through the
factory on a flow basis. At the other extreme the inexpensive class
C parts are ordered in large batches with generous processing times
and using a sealed-stock reordering system. Class B parts can be
ordered in medium-sized batches using stock-level recording operated
from an appropriate record of stock and outstanding orders.

10.2.7 Stock Records
Accurate stock records are essential for the satisfactory operation
of most ordering systems. If the recorded stock is too high, there
is a risk of running out of stock; if it is too low, parts will be
ordered unnecessarily. Stock records can be in the form of hand-
posted bin cards kept in the stores, machine-posted stock record
cards in an office, or numbers stored in a computer. In general, the
more remote the stock record is from the stock location, the more
difficult it is to ensure accuracy. Bin cards and stock record cards
provide an historical record of stock transactions on each part
thereby simplifying the detection of errors. This facility does not
exist with computerised stock records and considerable effort is

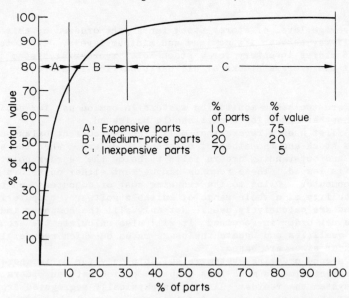

Figure 10.10   Distribution of the value of parts in a typical
               engineering product

needed to ensure that stock-movement input is correct. Computerised
systems have the advantage that multiple access to information on
stock levels can be obtained within seconds by means of visual display units (VDUs) in offices and stores.
   As a high level of stock-level accuracy is important for the successful operation of most reordering systems, some form of perpetual
inventory control should be used. This ensures that all stocks are
physically checked at regular intervals and any necessary adjustments
made to stock record balances. When errors are discovered, efforts
should be made to determine their cause so that their future occurrence can be minimised.

10.2.8 Issue of Materials
Material needed for a production order is normally released from
stores only on the presentation of a material requisition. This
requisition will specify the material and state the exact amount of
it to be issued. Any loss of material in production can be replaced
only on the presentation of the appropriate material-replacement
requisition.

10.2.9 Control of Parts in Process of Manufacture
Works Orders
Control is achieved in batch-production and jobbing-production by
some form of works order. A typical works order of the type that
accompanies the parts in the factory is shown in figure 10.11. This
document indicates the parts to be manufactured, the quantity required, the material to be used, and the sequence of manufacturing operations. The order also acts as a scheduling document and a record of
manufacturing achievement.

WORKS ORDER								
PART No. 345876		DESCRIPTION BRACKET				QUANTITY 11 000		
COMPLETION WEEK 36		ORDER No. 67541				MIN. QUANTITY 10 500		
MATERIAL SPECIFICATION MILD STEEL 2.5 mm x 40 mm BRIGHT STRIP						WEIGHT/1000 12.2 kg		
MINUS WEEK	OP. NO.	OPERATION	M/C GROUP	STD MIN/100	QTY PASSED	QTY REJECTED	DATE	INSP. STAMP
2	1	BLANK & PIERCE	PD2	M3	10 900	–	6 MAR.	⑫
1	2	FORM	P4	F2	10 700	200	10 MAR.	⑫
1	3	TUMBLE	B1	M3	10 700	–	12 MAR.	⑦
		PARTS STORE			10 700		14 MAR.	③

Figure 10.11 Works order showing variable and fixed information

Plant-loading

Additional control of work in progress can be achieved by some form
of plant-loading that allocates priorities to the processing of
orders. A basic rule for loading work is to give priority to orders
having the earliest priority date. If there is a choice of orders
having the same priority date, those with the shortest processing
time should be loaded first. It is suggested that a simple, rather
than an elaborate, form of shop-floor loading system should be used.
The system should be operated by production control staff in consultation with shop supervision.

Another aspect of plant-loading is the comparison of load with
capacity, often looking several weeks ahead. Computers can rapidly
produce regular load statements analysed by machine group. These
statements, which can be presented in bar-chart form, give early
warning of future load and enable work to be diverted from overloaded
to underloaded sections of the factory. Alternatively, if no internal spare capacity is available, work can be re-routed to subcontrac-
tors.

Progressing

A 'longstop' function is necessary with all production control systems
since some parts shortages are inevitable.

In many factories, progressing starts when sets of parts are physically collected in stores one week or more before they are required
for assembly. With computerised production control systems, the
labour-intensive pre-selection of parts may be avoided since progress-
chasers can be given a print-out of the late orders that are likely
to hold up the assembly operation.

If desired, a control file can be maintained showing the location of all work in progress. In many computerised systems, the movement of parts between factory operations and in and out of stores can be reported on-line to the computer from data-entry stations. The position of orders in the factory can be found by using VDUs connected to the computer.

10.3 LABOUR CONTROL

In jobbing-production and short-batch production, direct labour costs can be a major part of the cost of manufacture. To assist in the effective control of labour, some form of work measurement is required to provide reliable times for factory tasks. These times can be used in any type of production for the following purposes.

(1) To measure the performance of individuals and groups.
(2) To divide work fairly between those working in a group.
(3) To calculate the amount of labour needed for a given production programme.
(4) To reward effort by means of a wage incentive scheme.
(5) To estimate the labour cost of manufacturing operations.

For a production time to be meaningful, it must refer to a specific carefully defined operation, executed at a given level of performance.
There are four major methods of work measurement, each with its own particular field of application:

(1) Time study
(2) Synthesis
(3) Predetermined motion time systems
(4) Estimating

No method of work measurement is completely accurate. However, with competent and well-trained work study staff, acceptable time values can be obtained.
One difficulty found with most work measurement is drift. This is a long-term slippage caused by the easing of time values found difficult to attain and by social pressures lengthening the times allowed for new jobs.

10.3.1 Time Study
This technique was developed by C.E. Bedaux from the pioneering work of F.W. Taylor; it is widely used and has been employed in its present form since the 1920s.

Procedure when Making a Time Study
The steps that should be followed are summarised below.

(1) Obtain a stop-watch, the relevant operation layout, data recording forms, and a study board. The forms and watch are clipped to the board.
(2) After checking with the foreman, introduce yourself to the operator, put him at ease, and ask to watch the operation being performed.
(3) Ensure that the authorised method of production, as indicated by the operation layout, is being used.

(4) Check that the authorised method is the most appropriate and cannot be improved by the application of method study.

(5) Break the work cycle into elements. Elements are distinct parts of the work cycle which are separately timed. They are classified into either repetitive or occasional elements, depending on whether or not they occur in each cycle. The elements should be chosen to have distinct break-points and should not be so short that they are difficult to time. Manual and machine elements should be separately recorded.

(6) Note any relevant data concerning the study not recorded in the standard data; for example, operator's name and working conditions.

(7) Rate and time each element. It is usual to time about 30 cycles of work so that a representative sample of the work is observed. Rating is discussed in the next section.

(8) Calculate the basic time for the elements recorded in the study

$$\text{basic time} = \frac{\text{observed time} \times \text{observed rating}}{\text{standard rating}}$$

(9) Calculate the average basic time for each of the repetitive and occasional elements in the work cycle. The basic times for occasional elements should be divided by their frequency of occurrence.

(10) Add the appropriate relaxation allowances to the various average basic times. Relaxation allowances will vary considerably depending on the demands of the elements making up the work cycle. Tables of relaxation allowances are available in standard work study texts.

It may also be necessary to add a contingency allowance; these compensate for the legitimate and expected extra work and delays, such as machine adjustments, which occur but have not been included in the study.

The time now calculated is referred to as the standard time for the job. If management makes an addition to the standard time to increase the earnings, it is referred to as a policy allowance; the time for the job is then called an allowed time.

Rating

This is the assignment of a numerical value to an observed rate of working. A number of rating-scales are used; the two most popular in this country are the Bedaux 60/80 and the British Standard - the former takes 80 and the latter 100 as the value of standard rating. Standard rating is the rate at which an average qualified and motivated worker will work naturally at a specified task. When rating a time study, the observer assesses for each element the speed of working in relation to the job difficulty and then assigns a numerical value, usually in steps of five units. A standard performance is the average level of effort expected in an efficient factory. The distribution of performance expected from an unmotivated group paid by time-work only and from a well-motivated group is shown in figure 10.12. The non-standard 60/80 and the 0-100 British Standard rating scales are compared against a series of benchmarks in figure 10.13.

Since rating depends on the judgement and experience of the observer and since changes in working pace of less than ten per cent are difficult to detect, a high level of rating accuracy cannot be expected. Individuals vary greatly in their natural ability to rate accurately; however, training can improve both accuracy and consist-

ency. A reasonable level of rating accuracy is necessary, otherwise
tight and loose standard times will result, with ensuing labour
difficulties.

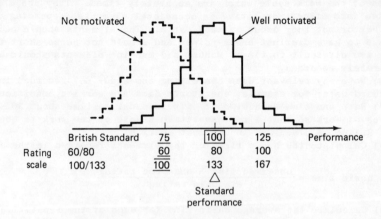

Figure 10.12   Distribution of operator performance

Rating Scale		Description of Pace	Walking Speed (mph)	Dealing 52 Playing Cards into 4 Piles (s)
BS 0/100	60/80			
50	40	slow	2	45
75	60	steady	3	30
100	80	brisk	4	22.5
125	100	fast	5	18
150	120	very fast	6	15

Figure 10.13   Rating benchmarks

10.3.2   Synthesis
This method of work measurement uses the elemental times obtained
from previous time studies, or from synthetic data, to establish
standard times. Synthesis reduces the cost of work measurement on
jobs containing a large proportion of common elements, such as certain machining and forming operations. Elemental times to be used in
synthesis should be unambiguously described and carefully chosen; if
the elements are to have a wide range of application they should be
reasonably basic. Where an elemental time is not available from
synthetic data, it may be possible to interpolate from closely related elemental times otherwise it will have to be obtained by time
study.

10.3.3   Predetermined Motion Time Systems (PMTS)
This method of work measurement uses time for basic human movement to
build up the standard time for a job. The elements used to build up
the time are much shorter than those used in synthesis and are of
universal, rather than specialised, application. PMTS overcomes many
of the disadvantages of time study; there is no timing, so stop-watch
errors are eliminated and the susceptibilities of workers who dislike

stop-watches are not offended. The PMTS times are issued at a defined level of performance, making rating unnecessary and avoiding
tight and loose times caused by rating errors. Although the elimination of rating and timing makes PMTS systems more difficult to criticise than time study, this does not mean that they are more accurate than good time study. The work method analysis must be thorough
and, since the tables themselves cannot be comprehensive, sound
judgement is needed in their use. This method of work measurement
has a wide and growing field of application.

There are many PMTS systems; the best-known are Work Factor and
Methods Time Measurement (MTM). MTM consists of MTM-1, MTM-2, and
MTM-3 and a number of systems specially designed for clerical and
maintenance work. All MTM times are expressed in Time Measurement
Units (TMU), where 1 TMU = 0.00001 hour = 0.036 second. No allow-
ances are included in MTM tables. The performance level at which the
times are issued is BS 83 and to convert TMUs into minutes at 100 BS
performance the TMUs should be divided by 2000. Basic motions in
MTM-1 are classified as Reach (R), Move (M), Turn (T), Apply Pressure
(AP), Grasp (G), Position (P), Release (RL), Disengage (D), Eye
Travel (ET) Eye Focus (EF), and various Body, Leg and Foot Movements.
The tables for these basic movements incorporate as appropriate the
TMU values for distance moved, job difficulty, force required, etc.;
there is also a table for guidance when there are simultaneous movements. The job being studied has first to be analysed into elements
so that the motion times can be extracted from the tables. For
example, reaching 10 inches to a tool in a fixed location would be
coded as R 10 A and from the Reach table this element is given a time
of 8.7 TMU (0.31 s). The total time for the job is found by adding
up the elemental times after allowances have been added.

The MTM-2 system is particularly useful to production engineers,
since it enables times for jobs to be determined while still in the
pre-production stage. A particular application is the balancing of
work between stations on an assembly line prior to installation.

MTM-3 is a further simplification intended for use in small-batch
work; here there is considerable variation in work method from cycle
to cycle. It can be more rapidly applied than MTM-2 but is less
accurate.

10.3.4 Estimating
Work measurement was originally developed for use in direct operator
incentive schemes when direct labour costs were a far larger proportion of product costs than they are today. Now in many industries,
indirect labour costs greatly exceed those of direct labour. Owing
to the variable nature of indirect work it has been difficult to
provide time standards against which individuals and departments can
be judged. However, a number of estimating techniques have been
developed; these provide approximate time standards for indirect work
and enable some control to be exercised over labour performance and
manning levels. Two methods of estimating are described.

Analytical Estimating
In this method of work measurement the job is analysed into elements
by the estimator. The elemental times are obtained from synthetic
values if these are available; if not, the estimator uses his job
knowledge and judgement. The integrity and experience of the estimator is a key factor in the success of analytical estimating. The

estimator should also be trained in method study so that the best
method of doing the job can be specified. This method of work
measurement is used in tool-making and maintenance work.

Variable Element Time Standards
Common elements can frequently be identified in indirect tasks; for
instance, storemen check the quantity of incoming or outgoing goods,
clear shortages, collect sets of parts, issue sets of parts, issue
individual parts, answer queries, deal with paperwork, etc. The
elements may not be performed by individual storemen in any particular order of frequency but they do comprise his daily work. If
times were available for these elements and there was a record of
work done by individual storemen, then time required for their work-
load could be calculated and compared with the attendance time. Precise time standards for work elements are not possible owing to their
variable work content; for example, the time taken will depend on the
bin location. Elements are timed on about ten occasions and roughly
rated using the BS scale in steps of 25 units. After adjusting the
observed times by the ratings, the basic elemental times are obtained
by averaging.

10.3.5 Labour Utilisation and Performance
Labour Utilisation
The labour utilisation of a group of operators is the ratio of hours
they have booked on production operations to their total attendance
time. A low figure of labour utilisation indicates a wasteful use of
labour. Some of the major causes of low direct labour utilisation
are the non-availability of work or machines and interruptions caused
by machine or tool breakdowns. A good supervisor can control labour
utilisation to a satisfactory level by adjusting his labour force to
meet the current work-load and rapidly switching operators to other
work if they are held up by temporary breakdowns or shortages. He
can also see that a sufficient number of setters are employed to ensure that operators are not delayed by waiting for machines to be
set. A high labour utilisation is possible, however, on overmanned
tasks since people tend to extend a light work-load to occupy fully
the time available (Parkinson's law). Overmanning can be detected by
measuring labour performance.

Labour Performance
Employee effort can be detected if operator performance is measured.
Operator performance is calculated from the following formula

$$\frac{\text{Standard hours earned on measured work}}{\text{Actual hours spent on measured work}} \times 100$$

A standard hour is one hour of work done at BS 100 performance.
The performance of individual employees can be calculated daily and
supplied to their supervisors. This task is facilitated if computers
are used to calculate and print out operator performance. Low performances, such as those below BS 75, should be taken up with individual operators by their supervisors. Factory managers should monitor the average performance of each production section. It is in
the company's interest to secure high operator performances since
this results in fewer employees and a lower investment in manufacturing equipment for a given level of output.

10.4 PROJECT MANAGEMENT

Complex one-off jobs such as constructing factory buildings, ship-building, and the manufacture of special-purpose machine tools can be extremely wasteful of capital, time, and manpower unless properly planned and adequately controlled. To exercise control, a number of charting techniques differing in detail, but basically similar in concept, have been evolved. The original American systems were developed in the 1950s under the titles PERT (Programme Evaluated Review Technique) and CPM (Critical Path Method).

A chart called a network diagram is drawn to show the interrelationship of all the tasks involved in the project. This provides the basis from which the total project time can be evaluated, together with the phasing of resources. Subsequently, costs can be minimised and resources reallocated to give a more effective distribution within the time span allowed for completion of the project. Activity networks of this type are of use, not only for planning the project and scheduling resources, but also as a master plan which can be updated and amended as work proceeds.

10.4.1 Network Conventions

The charting conventions adopted vary in detail between companies, but the basic conventions relating to network construction are in general use. BS 4335: 1972 'Glossary of Terms Used in Project Network Techniques' proposes standard terms to reduce confusion. Individual tasks, usually mental or physical work, are called activities and are indicated by arrows. In general, activities are drawn in time sequence from left to right; the length of the arrows is a matter of drafting convenience and does not indicate their duration. The terminations of activities are called nodes or events and are shown by circles. Those marking the start of activities are at the tail of the arrows and are known as tail events; similarly, those marking the completion of activities are known as head events. Events are numbered in a left-to-right sequence for identification purposes. In large networks it is sometimes convenient to leave gaps between the numbers allocated to groups of events to avoid the need for renumbering if the network is subsequently amended.

Precedence rules are built into the networks by means of the activities that terminate or commence at any event. Figure 10.14 illustrates a situation where commencement of activities M and N is dependent on the completion of activities K and L. A convenient method for showing network restrictions is the sign <, used for indicating precedence.

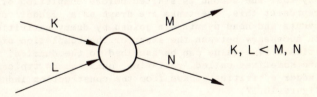

Figure 10.14 Activity-dependence diagram

In some cases these conventions are not sufficient to define a
network fully and dotted arrows known as dummy activities must be
included. Dummy activities usually have no time duration and require
no resources. They are of two types: identity dummies and logic
dummies. An identity dummy is used when two or more parallel activities have the same head and tail events. This type of dummy
enables activities to be allocated unique numbers, for example (1,2)
and (1,3), as shown in figure 10.15.

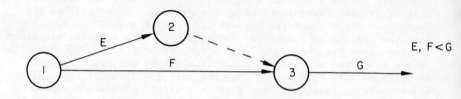

Figure 10.15 Use of identity dummy

Logic dummies are used when two chains of activities have a common
event, although they are in themselves wholly or partly independent
of each other. Figure 10.16 shows a situation where activity C cannot commence until activity A is completed whereas activity D cannot
commence until both A and B are completed.

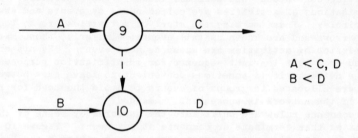

Figure 10.16 Use of logic dummy

Sometimes it is possible to schedule consecutive tasks in such a
way that the second is started before completion of the first. To
represent this, the tasks are drawn as a 'ladder' on which the tail
events and head events are joined by dummy activities. If a time-lag
is necessary between the start or the completion of the two activities a time value can be ascribed to the dummies, in which case they
are sometimes called 'real-time' dummies. A typical example of
ladder activities, drawn from the construction industry, is shown in
figure 10.17.

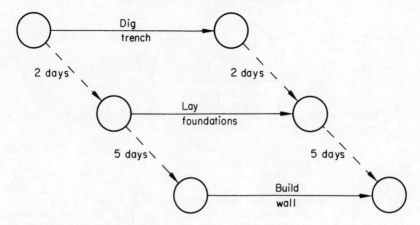

Figure 10.17 Ladder activities

10.4.2 Constructing the Network
Depending on the complexity of the network, the activities may be
simple operations or may themselves constitute networks of operations. In planning the network, the first step is to determine
the logical sequence and interdependence of the activities.

The next step is to estimate the time required to perform each
activity and from this information to calculate the earliest and
latest event times. A simple network shown in figure 10.18 will
serve to illustrate this procedure; the numbers associated with the
arrows represent activity times in days.

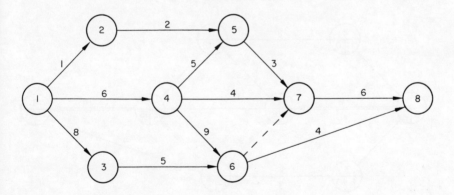

Figure 10.18 Network showing activity times

The earliest and latest event times are shown in the node circles
according to the convention illustrated in figure 10.19; other conventions may be used for illustrating event times.

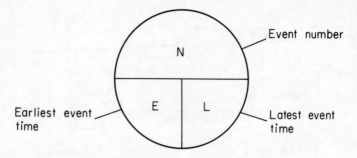

Figure 10.19   Event labels

If event 1 occurs at time zero, the earliest times of the other
events can be found by moving forward in the direction of the arrows
and summing the activity times. Where a node is entered by two or
more activities the earliest event time is the largest time value
obtained. For instance, the path times to reach event 6 are 15 days
if approached from event 4, or 13 days if approached from event 3.

The earliest time at which event 8 can be reached is seen to be 21
days. It is now possible to work backwards through the network to
find the latest event time that would permit the project to be completed by this date. In calculating latest event times when there
are two or more activities leaving a node, the earliest time is taken.
Thus the latest event time at node 6 would be day 15 if approached
from node 7 or day 17 if approached from node 8. The complete network is shown in figure 10.20 where the double lines show the critical path along which there is no spare time if the project is to be
completed by day 21. It is possible, although unlikely, for a network to contain two or more critical paths.

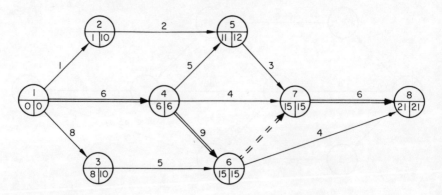

Figure 10.20   Network showing earliest and latest event times

10.4.3 Controlling the Project

When exercising control over time, resources, or cost, it is necessary to know what flexibility exists for scheduling non-critical
activities, flexibility being expressed in terms of float. The concept of float can be understood by considering activity 2-5 in the
network shown in figure 10.21.

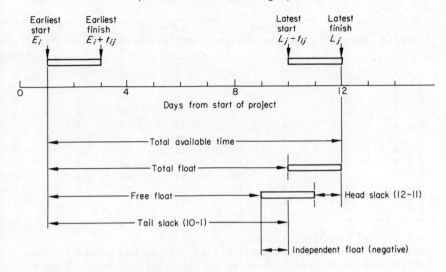

Figure 10.21 Floats and slacks (activity 2-5)

We can define an activity in terms of its starting and finishing nodes, i and j, so that

t_{ij} = time to perform activity

$\text{ES}_{ij} = E_i$ = earliest start to activity

$\text{EF}_{ij} = E_i + t_{ij}$ = earliest finish to activity

$\text{LS}_{ij} = L_j - t_{ij}$ = latest start to activity

$\text{LF}_{ij} = L_j$ = latest finish to activity

Total float is the amount of time by which an activity can be delayed without delaying the project completion date

$$\text{TF}_{ij} = L_j - E_i - t_{ij}$$

Slack is the difference between the earliest and latest event times. If the subsequent activity is scheduled at its earliest start time, the float will be reduced to give what is known as the free float

$$\text{FF}_{ij} = \text{TF}_{ij} - \text{head slack}$$

A further reduction of float will occur if the tail slack is absorbed due to a late start of the previous activity; this reduction of free float is known as independent float

$$\text{IF}_{ij} = \text{FF}_{ij} - \text{tail slack}$$

For the network being considered the floats are shown in the following table; the units are days.

Activity	Duration	Start		Finish		Floats		
		Earliest	Latest	Earliest	Latest	Total	Free	Independent
1 - 2	1	0	9	1	10	9	0	0
1 - 3	8	0	2	8	10	2	0	0
1 - 4	6	0	0	6	6	0	0	0
2 - 5	2	1	10	3	12	9	8	-1
3 - 6	5	8	10	13	15	2	2	0
4 - 5	5	6	7	11	12	1	0	0
4 - 6	9	6	6	15	15	0	0	0
4 - 7	4	6	11	10	15	5	5	5
5 - 7	3	11	12	14	15	1	1	0
6 - 8	4	15	17	19	21	2	2	2
7 - 8	6	15	15	21	21	0	0	0

10.4.4 Cost Reduction and Control

Some activities will occupy fixed time periods but others can occupy varying periods of time depending on the resources used. When resource levels are varied, the cost incurred may change. For instance, doubling the amount of labour on an activity does not necessarily halve the time taken; alternatively overtime working may reduce the time but increase the cost because of overtime payments. It is worth while assessing the cost/time sensitivity of activities so that advantage can be taken of relaxing the time on highly sensitive non-critical activities to reduce project cost. Similarly, if it is necessary to reduce the total project time, critical or near-critical activities that are the least expensive to shorten should be selected for adjustment.

If the schedule of activities is known, cumulative cost can be computed throughout the project. As the project proceeds, the recorded cumulative cost can be reassessed in relation to the progress to date, showing the historical cost variance incurred. Additionally, cost variations due to wage or material adjustments must also

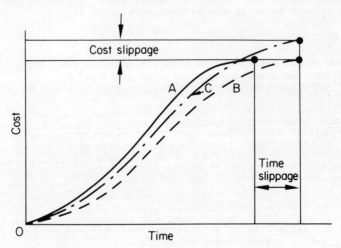

Figure 10.22   Cumulative cost/time curve for project

be allowed for when updating the subsequent activities. Typical
cumulative cost curves are shown in figure 10.22 where line A represents the original planned expenditure. If activity completion times
fall behind the original timetable and the lost time cannot be re-
gained, the project will have to be spread over a longer period, line
B. Expenditure on the project can also differ from that originally
estimated; the effect of both increased time and expenditure is shown
by line C.

10.4.5 Resource-scheduling

Two problems may arise with regard to use of resources in projects;
a resource may be limited to a maximum value or it may be desirable
to level its use to minimise peaks and troughs over a time period.
The first limitation can involve extending the project time so that
the limit on a particular resource is not exceeded.

Except in very simple cases it is not possible to specify a rule
that will optimise the use of resources. Generally a decision rule
is used which gives a reasonable schedule within the constraints imposed. Decision rules apply an index of priority to jobs competing
for a resource; jobs are then ranked in order of priority according
to their float, and resources are allocated in this order until they
are exhausted.

Resource-scheduling by manual manipulation is very tedious and
time-consuming for all but the simplest projects. With larger projects it can be achieved effectively only by controlling the network
on a computer. Computer packages for network problems are available;
these are capable of many alternative types of output showing resource and cost allocation in terms of time. For convenience of
project management, a Gantt-chart output can be used. Here the
individual activities are listed vertically and the relevant activity
times are shown as horizontal bars plotted against a base of time.

10.5 MAINTENANCE MANAGEMENT

The usual functions of the maintenance department are listed below.

(1) Keeping plant in such a condition that breakdowns are reduced
and work of an appropriate quality is produced.
(2) Repairing plant and equipment.
(3) Moving and installing plant and equipment.
(4) Maintaining buildings and their services.
(5) Factory cleaning and security.

The level of plant maintenance should be determined by the demands
of the manufacturing processes; in practice it is controlled by the
size of the maintenance budget.

Historically, maintenance has tended to be an emergency service
performed in an *ad hoc* fashion and largely concerned with getting
broken-down plant back into service as soon as possible. In recent
years there has been a growing awareness of the deficiencies in
maintenance and replacement policies. This has led to efforts to
integrate them into a more comprehensive approach which is known as
terotechnology. Terotechnology is concerned with the complete responsibility for physical assets from installation to replacement,
including those aspects of equipment design that affect durability
and ease of maintenance. Some aspects of terotechnology are illustrated in figure 10.23.

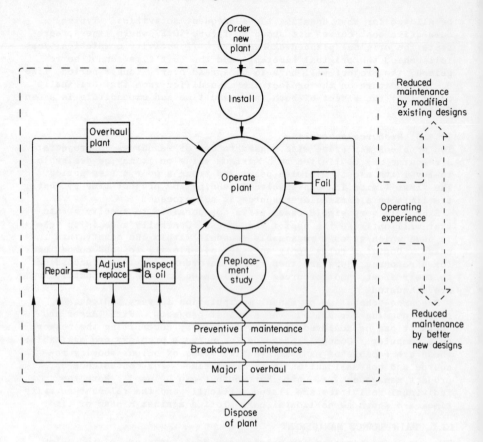

Figure 10.23 The terotechnology concept

The relative importance of the maintenance function depends on the
type of product and the production plant on which it is made. In
some process industries, maintenance costs can be in excess of ten
per cent of the cost of goods sold. There is no generally accepted
measure of maintenance efficiency; however, if the following param-
eters are calculated periodically they can provide useful trends.

(1) Annual maintenance cost/replacement value of plant maintained
(2) Annual maintenance cost/annual cost of goods sold
(3) Down time/total scheduled production time

10.5.1 Maintenance Policies

These vary from breakdown maintenance, in which the equipment is run
until it fails and is then repaired, to preventive maintenance where
an attempt is made to avoid breakdown by anticipating failure or wear
and making a timely examination, replacement, or adjustment. Preventive maintenance is usually considerably more expensive to operate
than breakdown maintenance; however, this additional expense must be
set against the savings resulting from minimising the random breakdown of plant. Unless there are overriding considerations of safety,

such as in aircraft maintenance, preventive maintenance should be
performed only when there is a net cost saving to the company.
Breakdown and preventive maintenance are not mutually exclusive.
Some form of preventive maintenance such as regular lubrication,
inspection, and adjustment is normally practised on almost all equipment, even when the main policy is one of breakdown maintenance.
Conversely, with comprehensive preventive maintenance, the unexpected
breakdown will still have to be repaired.

10.5.2 Preventive Maintenance

It is difficult to determine the correct level of preventive maintenance to apply. Often help can be obtained from the maintenance
schedules proposed by equipment manufacturers; however, a record of
the frequency of unscheduled breakdowns and their cost, both direct
and consequential, is really needed to institute and refine preventive maintenance programmes. This information is rarely available in
batch-production factories; in fact, in most of these factories
there is little information available even concerning the number of
hours individual machines have worked. As a result, preventive
maintenance schedules for batch-production plant are normally prepared on a calendar basis rather than on a running-time basis.

It would be a relatively simple matter to design a preventive maintenance system if equipment were to fail without variation after a
given usage. However, even simple items of equipment have fairly
widely spread probability density functions of failure.

To summarise, preventive maintenance should be considered when the
following conditions apply.

(1) The time interval between equipment breakdown can be predicted
with reasonable accuracy.

(2) The cost of preventive maintenance attention is less than the
repair cost when both costs include that of any lost production.

(3) Equipment failure is likely to disrupt subsequent production
operations or cause customer dissatisfaction.

(4) Injury could result from equipment breakdown.

10.5.3 Breakdown Maintenance

There are advantages in having this work centrally organised to ensure better utilisation and deployment of the work-force. Maintenance staff concerned with repair work should be trained in fault-
finding, be well-supplied with diagnostic equipment, and have makers'
handbooks and circuit diagrams readily available. On larger sites
the repair teams can be kept in touch with central control by two-way
radios.

Repair times can be reduced by having spare parts available in
stores rather than obtaining them from the equipment manufacturers.
When deciding on spares to be carried, some help can be obtained
from operating experience with similar equipment. The cost of
stocking spares and the time required to obtain them must be considered when deciding on the appropriate level of spares inventory.
If production equipment has been standardised, a reduction in the
total inventory of machine spares can be achieved.

10.6 MANAGEMENT INFORMATION SYSTEMS

The successful operation of a production system requires that oper-

ational and product data are stored in an accessible form and that
appropriate data are collected, processed, and distributed. Relevant, timely, and accurate information is required for decision-
making; this will vary from lists of overdue orders to be used by
progress-chasers and shop foremen to profitability and market-share
trends for senior management.

10.6.1 Computer-based Management Systems

In the 1950s, early computer installations were used to perform more
rapidly specific clerical tasks such as calculating the weekly pay-
roll or determining how many parts should be ordered to meet a production schedule. The ease with which computers generated data often
resulted in the production of vast quantities of tabulated information much of which was irrelevant to its recipients. The 1960s saw
companies beginning to abandon their application-by-application
approach and attempting to amalgamate individual applications into
more closely integrated management information systems. This reduced
data-collection costs by using common, rather than separate, data
files as well as avoiding the partial duplication of file data. The
period also saw an improvement in the speed of information retrieval
and the development of interactive systems which enabled managers to
'converse' with the computer.

Apart from the difficulties of designing or selecting an appropriate information system, a major system failure could produce far
more serious effects than a comparable failure of a simpler clerical
system. The inflexibility of many computerised systems can prove an
obstacle to management innovation, particularly at middle-management
levels. These limitations must be realised and allowances made when
designing information systems; every opportunity should be taken to
encourage management participation in the design and operation of
the system.

Design of Large Systems

The production system comprises a number of separate subsystems each
with its own information flow. An attempt has been made in figure
10.24 to show the main information flows necessary to sustain production. In the design of large systems, the data-processing capacity of the computer should be effectively used and, where possible,
data should be presented in such a way as to assist management in
decision-making. Rapid access to stored data should be provided, and
data should be maintained at an appropriate level of accuracy and
currency.

Data should be stored in convenient groupings and their duplication
in store avoided. To reduce the volume of information distributed,
the exception principle of reporting should be used where possible.

The data system is likely to be hierarchical with recognisable subsystems which can be further subdivided into modules. This arrangement provides flexibility since individual modules can be altered,
thereby retarding the obsolescence of the whole system. For example,
a stock control subsystem could consist of recording issues and
receipts, updating stock balances, predicting future demand, calculating reorder quantities, and determining whether or not an order
should be issued. If all these are treated as separate functional
modules, a new method of predicting demand could be used at some
future time, without modifying the whole subsystem.

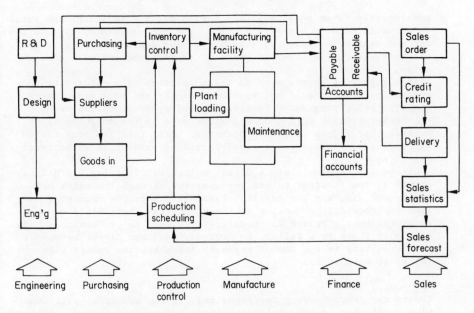

Figure 10.24   Information flow in manufacturing

Information subsystems usually consist of well-defined groupings
of modules the interaction of which remains largely within the subsystem. Subsystems, however, interact with others and information
is shared; for instance, a parts-ordering subsystem will use information from those concerned with product data and production scheduling.

Information Control

The organisation of information is a major task in the design of
management information systems. Information can come from outside
sources such as suppliers, customers, and government departments or
can be internally generated from management policy, product specification, and manufacturing data. Some information is of a transient
nature and once used can be discarded; other information must be retained in the system for subsequent analysis.

In designing a management information system, the main channels of
information flow should be identified; these will not necessarily
conform to formal organisation boundaries. Although there are information flows downwards and upwards in the organisation hierarchy,
heavy flows occur horizontally at supervision level between sections
and departments. Sometimes formal systems remain in operation long
beyond the time limit of their usefulness whereas informal systems
are used only because there is a need for them. Sources of information should be identified and classified; only when this has been
done and the mainstream of information flow has been established can
decisions be taken concerning the design of the system.

Activity Reporting

Activity reporting is expensive and a high level of accuracy is
difficult to maintain. Data can be processed either as batches or

continuously by an on-line system. With batch processing, the input
is accumulated over a period of, say, one day, sorted by type of message and then processed. On-line processing deals completely with
each message in the order in which it is received and with minimum
delay.

When reporting events such as the completion of a manufacturing
order, the event should be reported promptly and once only, the
computer being programmed to classify and cross-index the event.

A fast-declining method of reporting data to the computer is by
punching cards from original documents. With this indirect method
there is a time delay and a greater risk of error than with direct
reporting.

Data-entry terminals are a direct method of reporting. A display
screen at the terminal guides the operator through the entry procedure and displays the result. Transmitters can be connected on-
line to important production equipment so that their state can be
monitored by a computer and modifications made to production schedules when there is a failure to produce. Another direct method of
data-reporting is the use of magnetic ink character recognition
(MICR) as on cheques.

Accuracy of Data

Errors may produce wrong decisions and, if too numerous, will under-
mine confidence in the whole management information system. Although
it is impossible to eliminate all errors, particular attention should
be given to eradicating those that can result in costly mistakes.

It is possible to minimise errors if information is collected in
its simplest and briefest form; they can also be reduced if there is
minimum delay in reporting and the reporter is the person concerned
with the activity. With remote data-collection terminals it has been
found that errors increase exponentially with the number of variables
reported.

Errors of omission, that is, completely failing to send a piece of
information, are usually more difficult to detect than mistakes in
transmitted information. Format and range checks can be devised to
detect errors. Rules can also be formulated for particular instances; for example, physical stocks of parts cannot be negative.
Where possible, errors should be detected before any of the incorrect
data are processed. Preferably, the sender should be informed of his
mistake immediately by means of an error message on the display
screen of the terminal. This enables errors to be corrected at once
by the sender.

Data Organisation

In management information systems a vast amount of data is stored;
once stored the data must be easily located and retrieved. Information is stored as coded data elements, each of which should be
clearly identified. Three types of element are identifiable. Static
elements are those kept permanently on file until they are modified
or discarded; for example, the parts-listing of a product. Dynamic
elements form the second group; these indicate current plans or
status such as assembly schedules or work-in-progress levels. Thirdly, there are historic elements which are stored for record purposes;
for example, monthly sales for the past year.

All data should be organised so that logical access exists to each
item. The storage is in compatible groups called files; an example

of file organisation is shown in figure 10.25 where it is possible
(a) to find all parts required by a particular assembly and (b) to
discover in which assemblies a part is used.

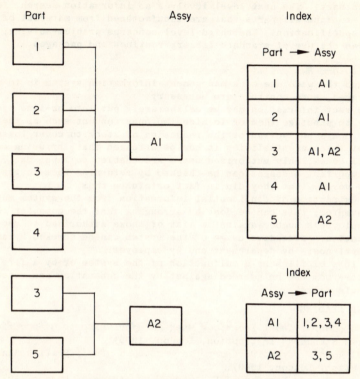

Figure 10.25 Simple product-structure indexing

Rather than duplicate data in a series of independent files, all
information on a particular subject should be contained in single-
source files, for example, combining into a single personnel file
information on staff addresses, qualification, present position,
etc., previously held on separate files. Single-source files reduce
the work of updating, the risk of error, and storage costs.

The idea of single-source files can be taken one stage further if a
data base is used to form a single comprehensive collection of company
data. Data elements require extensive organisation and indexing and
elaborate software to generate reports and interpret requests. The
high cost of the more complex systems should be balanced against the
greater utility given to the stored information.

Information Output
Printed output can be obtained from a central printer or from printers
situated at remote work-stations. The amount of information that a
manager has to review can be reduced considerably by the exception-
reporting of information that is of direct concern to him only.

Interactive systems enable employees to obtain specific information
directly from data banks; they also enable problems to be solved. The

main method of conversing with a computer is to use a keyboard input
with a visual display. Various levels of complexity exist, the lowest
being simple data retrieval such as the quantity of stock held of a
certain part. The next level involves an information search, for example, to find all parts that are manufactured from material of a
given specification. The third level concerns problem-solving; this
requires the use of standard library routines and packages.

Security and Reliability

Although the wide use of a management information system is to be
encouraged, some controls are necessary.

Incorrect information may be deliberately put into on-line systems
either to sabotage them or to hide improper conduct such as the booking of hours not worked or the reduction of stock to cover losses.
To meet these eventualities it can be arranged that, by using confidential codes, only authorised users are enabled to input data. Certain important messages can be checked by returning them to the send-
ers to verify that they did in fact originate them.

The extraction of confidential information from the system must also
be prevented; this can be done by arranging that the computer checks
the employee's number against a list of those authorised to receive the
information. Excessive usage of the system can be checked by analysing
retrieval costs by department and by employee.

The loss of files by a malfunction of the system or by a fire in the
computer-room can be guarded against by the generation and retention of
duplicate files.

FURTHER READING

BS 2564 *Control Chart Technique when Manufacturing to a Specification*
(British Standards Institution, London, 1955).
BS 5701 *Number Defective Chart for Quality Control* (British Standards
Institution, London, 1980).
Currie, R.M. and Faraday, J.E., *Work Study* (Pitman, London, 1977).
DEF 131A *Sampling Procedures and Tables for Inspection by Attributes*
(HMSO, London, 1967).
Dodge, H.F. and Romig, H.G., *Sampling Inspection Tables* (Wiley, New
York, 1959).
Hill, R.W. and Hillier, T.J., *Organisational Buying Behaviour*
(Macmillan, London, 1977).
Juran, J.M. and Gryna, F.M., *Quality Planning and Analysis* (McGraw-
Hill, New York, 1980).
Kelly, A. and Harris, M.J., *Management of Industrial Maintenance*
(Newnes Butterworth, London, 1978).
Lockyer, K.G., *Production Control in Practice* (Pitman, London, 1975).
Shave, M.J.R. and Bhaskar, K.N., *Computer Science Applied to Business
Systems* (Addison-Wesley, London, 1982).

Index